TIMELINE

Science & Technology

Peter Goes

GECKO PRESS

Homo heidelbergensis, an extinct ancestor of humans, was the first to build shelters and to regularly hunt with spears. They lived 700,000 to 300,000 years ago.

The 44,000-year-old Lebombo Bone, found in southern Africa, is a baboon's leg bone with notches cut into it — maybe for use as a calendar.

The oldest discovered musical instrument is a 42,000-year-old flute.

A microlith is a small flint tool less than 5 cm / 2 in long.

From a hammer stone to an arrowhead made of flint, the development of stone tools was a slow evolution that began 2.6 million years ago.

The boomerang was depicted in a 25,000-year-old Australian rock painting.

A spear thrower allowed hunters to cast a spear further and more powerfully.

People have enjoyed sitting around campfires for thousands of years.

Bread is one of the oldest prepared foods.

The use of boats probably dates back 800,000 years.

It's hard to be sure when people first learned to make fire. After a forest fire, they would have carefully tended a flame in a sheltered place. Later on, people made fire by rubbing dry sticks together or using hot sparks struck from stone tools.

Wooden spear points are stronger if hardened in fire.

Pestle and mortar.

The Old Stone Age or Paleolithic

The Stone Age was the time of stone tools. Tools of wood and bone were also used, but because these materials decay easily, hardly anything of them remains. The Stone Age can be divided into three stages: the Old Stone Age (Paleolithic), the Middle Stone Age (Mesolithic) and the New Stone Age (Neolithic). The Old Stone Age is the first period in prehistory when people used tools. It began about 3.3 million years ago and finished 12,500 years ago with the end of the last ice age. Until about 10,000 years ago, people lived nomadically as hunters and gatherers. Small groups moved from place to place looking for enough to eat. Hunter-gatherers left Africa in search of new sources of food. They eventually settled in every continent except Antarctica.

Using bone needles, people made warm clothes from bark and animal skins.

Archeological remains, such as animal bones with embedded flint fragments, suggest that bows and arrows were in use up to 18,000 years ago. However, firm evidence goes back only 10,000 years.

Dogs were the first animals to be domesticated.

Many tools were made from antlers.

Drilling ice cores can teach us about the past. Every fall of snow contained air, which became trapped. Ice cores from varying depths reveal information about Earth's climate in different periods.

Dolmens (stone monuments) from the Stone Age are found in Marayur, India.

The oldest complete pots were discovered in Japan, used by hunter-gatherers 16,000 years ago. In most other places, pottery was not used until the development of agriculture.

Traces of seeds can be found in pots. Palynology, the study of pollen, can tell us about how our ancestors lived and farmed.

The Middle Stone Age or Mesolithic

The Middle Stone Age was a transitional period between the last ice age, which signalled the end of the Old Stone Age, and the beginning of the farming society that heralded the start of the New Stone Age. Because agriculture developed at different times and in different regions of the world, there is no specific end date for the Middle Stone Age. Even within a region, agriculture developed at different times—for instance, in southeastern Europe around 7000 BCE, in central Europe around 5500 BCE, and in northern Europe around 4000 BCE. Furthermore, in some regions there was no Middle Stone Age period. One example is the ancient Near East, where agriculture developed as early as around 9000 BCE.

This mammoth's tusk (from Gontsy, Ukraine, is nearly 15,000 years old. The lunar cycle engraved into it is one of the earliest examples of time measurement.

Stone lamps were found in France's Lascaux caves, which contain wall paintings estimated to be 17,000 years old.

As many as 200 stone pillars, up to 6 m / 20 ft tall, stand in 20 circles in southeastern Turkey. This stone monument, nearly 10,000 years old, is the oldest megalith so far discovered. The word megalith, which comes from Ancient Greek, means "large stone."

The climate became warmer and the ice caps melted. New land appeared from under the ice. At the same time, the sea level rose and low-lying lands disappeared. As the landscape changed, islands appeared, and continents that had been accessible by foot became separated. The map of the world as we know it today took shape.

The New Stone Age or Neolithic

Following the end of the last ice age, Earth's climate warmed up. What came next was a period of socio-economic and technological transformation, which we call the New Stone Age or Neolithic. Depending on the region, it began around 12,000–5000 BCE and lasted until 3000–1200 BCE. People discovered how to grow crops—helped by a warming climate—and domesticate animals. Small settlements surrounded by farmland sprang up in various places. People could build up supplies of food, becoming less dependent on hunting and gathering. They fashioned agricultural tools, and their settlements expanded into towns.

Pitch was used as a glue to secure axe and arrow heads to their shafts.

Bees were semi-domesticated. They were carefully selected to reduce both swarming and stinging. Bees are the only insects that produce food people eat. As well as honey, honeybees yield beeswax, royal jelly and propolis ("bee glue"—which may have use in human medicines).

The invention of the wheel was enormously important in the history of transportation, allowing people to move large loads.

After dogs, sheep were the first animals domesticated by humans, 13,000–10,000 years ago. To begin with, sheep were kept only for their meat, milk and skin. Selecting sheep for their wool began around 6000 BCE, while the earliest clothes made from wool date from 2000–3000 years later.

Some small settlements grew into wealthy Neolithic towns. Jericho in the Jordan Valley, for example, was by 7350 BCE already a walled city with wells and a population close to 3000 people.

Around 4000–3000 BCE, people began to tame and ride horses.

Copper was first used only in its pure state.

The body of a man who lived 5300 years ago was found in a melting glacier in Europe in 1991. Because the ice man, Ötzi, was so well preserved, we have learned a lot about the food, equipment and clothes from the Copper Age.

People began to extract copper from ore—rocks that contain minerals or metals. The rocks are heated in an oven until the metal becomes liquid. This process is called smelting.

The Copper Age or Chalcolithic

The Copper Age is the final period of the New Stone Age. Copper was probably the first metal used in prehistory. The gleaming metal was used, among other things, to make decorations, tools and sculptures.

Urban revolution

Agriculture caused great changes in the way people lived together. Forests were cleared so land could be cultivated. People discovered how to grow different kinds of crops and how to preserve the harvest. They invented new technologies, such as better farm tools and irrigation systems. Intensive agriculture ensured the availability of more, and better, food. More and more villages sprang up, bringing with them the need for social and political organization. The villages grew in population and size, gradually developing into towns.

A shadoof is a simple crane for lifting water. One of the earliest tools for irrigation, it is still used today.

Improvements in farming happened through many centuries of careful selection by farmers of their breeding stock. Farmers selected the best seeds to strengthen brittle ears of wheat and bred from sheep with thicker fleeces.

Irrigation can cause salt to build up in the soil, making it less fertile. To prevent farm fields from getting saltier each year, the Mesopotamians rotated which fields they farmed.

Due to its large mass, a flywheel needs little energy to keep it in motion once it gets started. The principle dates back to the Neolithic spindle, the potter's wheel and the round grindstones of antiquity.

Complex systems were designed to irrigate the land effectively. The Mesopotamians dug canals to flood their fields. The largest canals were more than 15 m / 9 mi wide.

The Bedolina Map is a rock engraving in northern Italy. It is one of the oldest topographic maps. Fields, houses, paths and people are pictured on it.

Stonehenge may have served as an astronomical observatory and a site for religious rites. It took decades of observation—and spectacular heavy-lifting skills—to accurately align the gigantic stones with the positions of the celestial bodies.

Stonehenge is a megalithic monument on Salisbury Plain in southern England. It was built between 3100 and 1600 BCE.

Discovered in Germany, the bronze and gold Nebra sky disk is the size of a dinner plate. Crafted more than 3600 years ago, it is the oldest known representation of the night sky.

The Bronze Age

An important discovery was made in the Middle East and in Asia around 3500 BCE: if you add a small amount of tin to copper, you make bronze. Bronze was harder than other metals of the time so it was highly prized. From 3000 BCE, knowledge of making bronze spread to Europe. People made bronze tools, weapons, building materials and decorative objects.

The first civilizations

The first kingdoms and early civilizations developed at about the same time in various parts of the world. Thanks to a warmer climate and the use of irrigation, more people were being fed. In this way, large and wealthy groups developed. The challenge of organizing the new, large populations led to strict systems of leadership—and to laws and armies. Writing was invented and refined to codify these complex structures. Large numbers of people worked together, willingly or not, to create huge monuments, often by a king's command. We have more evidence of the presence of organized religions beginning in this era.

Civilizations developed in many places, including the north of Mesopotamia from about 4000 BCE, Peru from about 3500 BCE, the Indus Valley from about 3300 BCE, Egypt from about 3000 BCE and China from about 2000 BCE.

Mesopotamia

The land between the Tigris and the Euphrates, rivers in the Middle East, was extremely fertile. A number of villages were settled around 6000 BCE. The inhabitants drained marshes and irrigated the land. An organized community developed: people engaged in trade and there were potteries and tanneries. In 3500 BCE the Sumerians conquered part of the land; in the Sumerian kingdom, great walled cities arose, such as Uruk, Jemdet Nasr and Ur, each with 50,000 to 200,000 inhabitants. City-states and kingdoms evolved from these, with legal and political structures. During the following thousands of years a number of those kingdoms would come to dominate the area.

King Ur-Nammu built an enormous ziggurat, sacred to the Sumerian moon god Nanna.

A kuphar is a small round boat, made of willow branches and animal skins.

Our present-day division of an hour into 60 minutes, a year into 12 months and a circle into 360 degrees derives from the Sumerian year and Mesopotamian notation, which had 60 as its basis.

A medical manual from ancient Babylon described the symptoms of all kinds of illnesses and offered remedies too.

The oldest pictographic script was invented in Sumer around 3300 BCE. Later, it became cuneiform—an alphabet of wedge-shaped marks.

Cylinder seal.

Before the development of writing, people used stones to keep a tally of goods. These were the first delivery notes. The stones were sometimes sealed in earthenware containers, which were marked to indicate the contents. Later, people used clay tablets.

The first archives and libraries.

The Norte Chico civilization

At the height of the last ice age, people from Asia migrated into the Americas. They represented various strands and cultures, and from them great civilizations arose. One of the earliest of these was the Norte Chico, or Caral, civilization. This society was based around 30 or so settlements—most of them beside the Huaura, Fortaleza, Pavivilca and Supe rivers—in the Norte Chico region on the north coast of Peru. The civilization arose in 3500 BCE and declined around 1800 BCE; at its height, Chico Norte was one of the world's most densely populated regions. Bandurria and Caral are probably the oldest known towns in the Americas.

Before the domestication of the llama enabled the use of its dung as fertilizer, early inhabitants of Peru collected guano—the droppings of seabirds.

Caral was an important trading hub for an area that stretched from the sea to the rainforests of the Andes.

In one temple, archaeologists found flutes made from the bones of condors, pelicans, deer and llamas. No ceramics were found so it seems that pottery had not yet been invented here.

Beans

Calabashes

No weapons or defensive walls were found in Caral. Historians conclude that the society was a peaceful one.

Fishing nets were knotted with cotton.

Reed fishing boat.

Llamas were domesticated in the Andes mountains.
They not only supplied wool, meat and manure,
but were also used to carry things.

The Norte Chico sites are noted for
their monumental structures and
complex irrigation systems.

A quipu was used to keep track of information and to communicate. By using a great
variety of colors, cords and sometimes hundreds of different knots, a quipu could
pass on information, data, stories and poems.

Blowpipe

Mohenjo-daro and Harappa were the largest cities. At its height, Mohenjo-daro spread over 3 km² / 300 ha and was home to 40,000 inhabitants.

Many images from Harappan culture—on small stone seals, for instance—show a half-bull, half-zebra animal, which was possibly a deity.

In the past, adults were lactose-intolerant. Unlike babies, they could not digest the milk sugar lactose so milk was mostly heated and fermented into yogurt and cheese.

The Indus Valley civilization

The Indus Valley civilization, or Harappan Culture, flourished in the northwest of South Asia, mainly in the basin of the Indus River. The precisely planned settlements—some of which grew to be huge cities—featured houses of brick, with sophisticated sanitation and drainage systems, but no temples, so far as we can tell. First forming around 3300 BCE and reaching its zenith a millennium later, Harappan society was made up from a cosmopolitan mix of races, and many of its city folk were traders who relied on boats and bullock carts to carry goods, such as worked metal pieces, ceramics, beads and cloth, perhaps as far as Egypt and Afghanistan. A shortage of drinking water, not enough fertile ground or a decrease in trade are all possible explanations for the population's surprisingly rapid decline in 1900–1700 BCE.

Approximately 1500 local archaeological sites are known, of which nearly 500 have been excavated over the last couple of centuries.

As well as plying the rivers, boats could have reached up into Mesopotamia via the Persian Gulf. A large shipyard on the coast at Lothal points to seafaring, though not all historians agree.

The Indus or Harappan script contained at least 400 different characters, and it has not yet been deciphered. About 4000 samples of text have been found, mostly in the form of small, square stamp seals made from stone, metal or bone.

People made fire with a fire drill or bow.

According to Egyptian mythology, the god Thoth was the inventor of writing and an authority on science and magic. He passed his knowledge on to mortals.

The Egyptians sometimes used their obelisks as sundials, observing shadows to determine the longest and shortest days of the year.

The pyramids of Giza were built by thousands of workers around 2613–2494 BCE. For some 3800 years, the largest, the Pyramid of Khufu, was the tallest human-made structure on the planet.

The Egyptians often erected two obelisks by their temples, one on either side of the entrance.

From 1500 BCE the Egyptians made decorations and bottles from glass.

From around 3000 BCE, Egyptians wrote on papyrus, made from the plant of the same name.

Water clock.

An Egyptian farmer used a wooden sickle lined with flint teeth to harvest grain. The ears were cut, while the stalks remained as feed for livestock.

Irrigation canals, ditches and sluices.

Vegetable gum, soot and beeswax were used to make a black ink.

Hieroglyphics, the script of Ancient Egypt, served as the basis for the Phoenician alphabet, from which later alphabets, such as Hebrew, Greek and Latin, were derived.

Ancient Egyptians cleaned their teeth with toothpaste made with burnt eggshells, ground ox hooves or peppercorns.

Goldsmiths and silversmiths made richly worked necklaces, bracelets and small utensils, often decorated with precious stones.

With the discovery of the Rosetta Stone, French linguist Jean-François Champollion (1790–1832) was able to decipher Egyptian hieroglyphics. The stone carried the same text in three languages: Egyptian hieroglyphics and demotic script, and Greek.

The Egyptians played an important part in the development of seafaring, including the use of lighthouses and pilots.

The Egyptians used a calendar of 365 days, approximately a solar year.

Copper scissors.

Metal mirror.

Wood lathe.

Egyptian archers rode in chariots rather than on horseback. That was a significant advance since the chariots provided a more stable platform for the archers and could also carry more arrows.

Ancient Egypt

The giant-scale temples and monuments of Ancient Egypt were built for deities or the all-powerful pharaohs. The Great Pyramid of Giza, the temple complex in Karnak or the Colossi of Memnon are among the most awe-inspiring examples—but everywhere in Egypt there are huge structures, thousands of years old, which have people wondering about their construction. Many answers can be found in the Ancient Egyptians' inscriptions, texts, murals, grave inscriptions, art and tools, which reveal their extraordinary understanding of science and technology. In addition to their enormous building works, the Ancient Egyptians made advances into nearly every area of life, from making simple household items and brewing beer to agriculture, medicine, astronomy, art and literature. The main eras of Egyptian civilization were the Old Kingdom (about 2700 to 2000 BCE), the Middle Kingdom (2000–1800 BCE) and the New Kingdom (about 1550 to 1050 BCE).

The first Chinese dynasty

In legend, the first Chinese dynasty was the Xia Dynasty (about 2070–1600 BCE), when the founder Yu the Great is said to have controlled the frequent flooding of the Yellow River by building dams and canals, creating fertile farming land. Next came the Shang (probably 1600–1046 BCE), the first Chinese dynasty for which there is archaeological evidence. Chinese inventions and discoveries cover every area of science and technology. Among other things, the compass, gunpowder, paper, printing, bronze forging, the seismograph, the crossbow, the iron plow, the wheelbarrow and the stern rudder were all Chinese inventions.

Early Chinese wrote on bamboo or on strips of wood, which they assembled into book form.

The first references to a candle clock can be found in a 2500-year-old Chinese poem. Later, candle clocks came to be used around the world.

Chinese farmers used a paddle wheel to irrigate their fields.

Printing began in China. Pictures and text were carved into wooden blocks and printed onto silk or paper. The oldest print dates back to 220 CE.

In the year 105 CE, Cai Lun (50–121 CE) invented a better kind of paper, made from bamboo, bark and scraps of silk. He is said to have watched wasps making their nests from a pulp of chewed wood.

Compass made from an iron ore pointer on a brass plate.

Crossbow.

Chinese cattle shoulder blades and tortoise shells retain the oldest known traces of primitive writing dating back 8600 years. Questions about the future were written on them, after which they were broken up with a red-hot poker. People used the shards to try to predict the future. These objects were called oracle bones.

From the 7th century BCE, various peoples in China erected defensive walls. Increasingly, the walls became linked, and, as the years passed, more sections were added. At its longest, the Great Wall of China measured 21,000 km / 13,000 mi. Today it is about 6000 km / 3500 mi in length.

The Chinese academic Zhang Heng (78–139 CE) was the inventor, among other things, of the seismoscope—an ornate device whose complex and sensitive mechanism could detect earthquakes.

Seed drill.

In the 1st century BCE the Chinese put a curved blade on their plows to turn the soil and smother weeds as they made furrows for seeds.

The early natural philosophers asked themselves big questions, such as "How did the world begin?," without seeking answers in mythological explanations.

Anaximander (c.610–c.546 BCE), a pupil of Thales, wrote on many topics, including geography, cosmology, astronomy and the origins of the universe. He also published one of the first maps of the world.

Pythagoras (c.570–c.500 BCE) and his later followers combined science and natural philosophy. His name was applied to the best-known mathematical theorem, even though he is unlikely to have been the first to discover or use it.

Thales of Miletus was said to have predicted the solar eclipse of 585. The eclipse took place during the Battle of Halys, frightening the soldiers so much that they called a truce.

The Diolkos track crossed an isthmus between the Ionian and Aegean seas. Grooves along its paved roadway suggest it served as a kind of tramway for transporting ships. It was in use from 600 BCE until the middle of the 1st century CE.

Many tales grew around Thales. He is said to have averted war between the Milesians and Lydians simply by offering them some good advice, and to have helped the soldiers of King Croesus invade Persia by showing them how to divert a river that obstructed them.

Empedocles (c.494–c.434 BCE) believed that everything came from (and eventually returned to) four elements: water, fire, air and earth, upon which the divine forces of Love and Strife acted.

The Iron Age

For the ancient Near East civilizations, the Bronze Age collapsed fairly suddenly in a wave of regional conflicts around 1200 BCE. In place of bronze came iron. This harder metal was better suited to making tools and weapons, and its ore is more readily available in nature than copper or tin. However, smelting iron requires higher temperatures than for copper, so people had to build better ovens. The technology probably dates from around 1200 BCE in Greece and the Near East, and within seven centuries had spread to northern Europe.

The philosopher, astronomer and mathematician Democritus of Abdera (c.460–c.370 BCE) believed that the universe consisted of emptiness and small, moving particles that he called "atoms." He experimented with bread, cutting pieces in half then in half again, asking himself what he would end up with if he continued, leading him to the concept of an atom (from the Greek atomos, "uncuttable"): an indivisible particle.

The philosopher Socrates (c.470–399 BCE) was renowned as a brilliant thinker and teacher. He used the Socratic method, a specific way of asking questions to develop ideas and reach truths, notably on issues of right and wrong.

The Athenian philosopher Plato (c.428–c.348 BCE) founded the Platonic school of thinking and the Academy, the Western world's first institution of higher learning. Not only an outstanding thinker, Plato was also a great writer.

Socrates left no writings of his own, but the works of his student Plato contain many accounts of his discussions and teachings.

Together with his teacher Socrates and his most famous student, Aristotle, Plato is a critical figure in the history of Ancient Greek and Western philosophy.

The Greek doctor Hippocrates (c.460–370 BCE) is famous as the founder of Western medicine. He put forward a diagnosis and treatment after making detailed observations. (This principle had already been followed in India and China for several thousands of years.) Before Hippocrates, people generally believed that illnesses were caused by the gods or angry spirits.

The Greek historian Herodotus (484–425 BCE) wrote the first history book, the story of the Graeco-Persian Wars. His writings remain an important source of information on life in Ancient Greece.

The Ancient Greeks

New forms of civilization developed along the coasts of the Aegean Sea. Instead of kingdoms where power was centralized, various independent city-states held sway. Classical Greek culture, based around one such city-state—Athens—reached its high point in the 5th century BCE. The Ancient Greeks observed the earth and heavens and studied humankind. They developed a unique brand of philosophy as a way of understanding their world without recourse to religion, myth or magic, and they founded the principles of Western democracy. Their fascination with mathematics led to many immediate practical uses, and many of their discoveries and inventions are still used today.

Aristotle (384–322 BCE) was one of the greatest thinkers of all time. His writings are so extensive and diverse that they could almost be called an encyclopedia of Ancient Greek knowledge. Aristotle's work influences nearly every aspect of modern-day thinking.

At the age of seventeen, Aristotle moved to Athens, where he became a pupil of Plato and other famous Greek philosophers. He studied at Plato's school, the Academy, until Plato's death.

Alexander gave his own name to many of the cities he founded. One was Alexandria in Egypt, home to the most famous library in the ancient world. Established in the 3rd century BCE, the library formed part of the Mouseion, the temple of the nine Muses, the Greek goddesses of the arts and sciences.

After Aristotle left Athens, he spent time in various Greek cities and also in Macedonia, where he published many of his writings and taught the young Alexander the Great and Ptolemy (the ruler-to-be of Egypt).

The odometer measures distances by counting the number of turns of a wheel. It may have been used to measure the journeys of Alexander.

Returning to Athens around 335 BCE, Aristotle established a school, the Lyceum. He encouraged his students to conduct research, and he built up a library of some 10,000 papyrus scrolls.

During his time with Aristotle, Alexander developed a great interest in science, and his army employed botanists and scientists to study the lands he conquered.

The Hydraulic Wheel of Perachora, dating from the 3rd century BCE, is the oldest-known pumping mechanism in Europe.

The Hellenistic period

In 338 BCE, Philip II of Macedonia (382–336 BCE) incorporated the Greek city-states into his kingdom. Philip's son, Alexander the Great (356–323 BCE), made one of the most remarkable journeys of ancient times, staking his claim over an empire stretching from the Ionian Sea all the way to the Himalayas. Alexander's invasion of Persia in 334 BCE marked the start of a three-century golden era in which Greek culture spread and flourished. This Hellenistic era ended when the Romans conquered Greece and the Ancient Near East in 30 BCE. Greek culture would become one of the most important foundations of Western society.

Hypatia (c.350/370–415 CE) was a philosopher and scientist, and one of the first known female mathematicians. She lived in Alexandria, where she was known as a wise teacher. Sadly, she was murdered for giving advice during a political conflict, and her death marked the end of Hellenistic influence in the region.

Hero of Alexandria (c.10–c.70 CE) was a scientist and engineer. His books are a treasure trove of information about science and technology in Ancient Egypt, Babylon and the Graeco-Roman world.

Built in the 3rd century BCE, the Pharos of Alexandria was one of the Seven Wonders of the Ancient World. This lighthouse was for many centuries one of the world's tallest structures.

Archimedes (c.287–c.212 BCE) was a Greek physicist, engineer, inventor and astronomer and one of the greatest scientists of all time. He worked out the value of pi, drew up the principles of hydrostatics, explained how levers worked, designed siege machines and solar heat rays, and much more.

Hero was a brilliant inventor and experimenter. He produced one of the earliest descriptions of an aeolipile, a form of steam engine that rotated by "jet power."

The Archimedes' screw was a useful pump for lifting water.

Recovered from a shipwreck off Greece in 1901, the Antikythera mechanism is, in effect, a 2100-year-old analog computer. A clockwork device with more than 30 bronze gear wheels, it enabled people to predict recurring events, such as solar eclipses, many years in advance. This level of technology was then lost, reappearing much later in 14th-century clocks.

Screw presses were a Roman invention, used mainly in the production of wine and olive oil.

From at least the 2nd century BCE, the Romans were using concrete in structures such as bridges and domes.

Roman cement was stronger than ours is today, thanks to the addition of volcanic ash: on coastal structures, this reacted with lime and seawater, hardening over time.

Aqueducts brought water to the towns and villages.

Boards coated with beeswax were used to write short texts or teach children to write, using a stylus. You then wiped the wax with a flat-edged tool to "make a clean slate", and reuse the board.

Vaulting, domes and rounded arches were elements of Roman architecture (adapted in medieval times as the Romanesque style).

The Romans laid highways throughout their empire to enable rapid troop movements and for transport.

In towns, surplus water was used to flush waste through underground sewers.

Cranes, tackles and pulleys, levers and screws—and the many applications that utilized them—were perfected by the Romans.

The Romans imported the ancient Syrian technique of glass-blowing into their empire. This quick method of manufacture meant glass was no longer a luxury product for the wealthy.

The scroll and the wax tablet replaced the first books, made of papyrus or parchment (from animal skins). Scrolls were still used for judicial processes and accounts until the late middle ages.

The chorobates was a mounted water trough (or sometimes a frame with two plumb lines). It was used to check that a construction site was flat—which was particularly important for aqueducts.

Public toilets were built all over the empire. They were stone benches over a channel of running water. The shape of the seat openings allowed people to clean themselves with a wet sponge on a stick.

From 59 BCE, news reports, carved onto tablets, were spread by the Acta Diurna, Rome's official daily "paper."

The hypocaust was an early central heating system, using an underfloor fire to warm the air beneath floors and baths.

The **Roman Empire**

The Roman Empire, which began in 753 BCE, gave the world an extraordinary range of new technology—mostly borrowed from earlier Greek science. Roman scientists did little of their own theoretical investigation, instead seeking practical applications for the ancient knowledge. For example, they used physics to calculate the power of catapults, biology to maximize land use, and mathematics to design domes and arches. Funding for these projects often came from wealthy investors.

Byzantine medicine continued to build on Hellenistic foundations. Medical books were written and hospitals were built for civilians for the first time. (Military hospitals already existed.)

At first, Byzantine wealth depended largely on grain exports from Egypt, which was then part of the empire. When Islamic forces of the Rashidun Caliphate captured Alexandria in 641, the grain trade suffered a huge decline.

Veterinary science.

John Philoponus (490–c.570) was also a pioneer of the theory of impetus: this addressed the movement of objects against gravity (such as a ball thrown into the air) and led to ideas about momentum and acceleration in classical mechanics.

The Byzantine Empire

In 395 CE the Roman Empire split into two parts: east and west. In 476 the Western Roman Empire fell, but the Eastern Roman—or Byzantine—Empire remained intact until it was invaded by the Ottoman Turks, roughly a thousand years later. At the start of the middle ages, the Byzantine Empire, with Constantinople (modern Istanbul) as its new capital, was the leading place of learning in the region, with Alexandria as its second city. Byzantine scholarship was a vital link between classical Greek and Roman learning, the Islamic world and the Italian Renaissance. Initially it provided the Islamic world with old and early medieval Greek texts about astronomy, mathematics and philosophy. Later, when the Caliphate and other Islamic cultures had the scholarly upper hand, the flow went the other way: Byzantine scholars translated Islamic scientific texts into medieval Greek.

A pendentive is an architectural feature that allows a dome to be placed over a rectangular structure (and thus to "square a circle"). The Romans had tried building with pendentives, but the problem was solved only during the completion of the Hagia Sophia in 537.

Emperor Theodosius II (401–450) founded the imperial University of Constantinople in 425.

The architects Isidore of Miletus and Anthemius of Tralles used complex mathematical formulas to design the cathedral (later a mosque) of Hagia Sophia. This engineering masterpiece would remain the world's largest cathedral for a thousand years. Today it is a museum.

Learning was widespread, sometimes even at village level. Girls went to school for around three years—not long, but in other places at that time, girls were not able to attend school at all.

The historian and theologian Nicephorus Gregoras (c.1295–1360) wrote a 37-volume history of the Byzantine Empire. He also wrote biographies, and covered topics as diverse as astronomy, dreams, funeral speeches and philosophy.

"Greek fire" (liquid fire) was used rather like a flamethrower to sink enemy ships in naval battles. Its ingredients were a closely guarded state secret and remain a mystery even today.

Based at his observatory in Iran, the Persian scientist and philosopher Nasir al-Din al-Tusi (1201–74) wrote more than 60 books on subjects ranging from astronomy and trigonometry to philosophy.

Chinese paper was in plentiful supply in the caliphate, following the creation of paper mills. After the invention of printing, books were to be found everywhere.

The House of Wisdom was the intellectual heart of medieval Baghdad. It was a library that quickly became a university where everyone who wanted to learn was welcome.

Hasan ibn al-Haytham (c.965–c.1040), also known as Alhazen, was the first to describe how light enters our eyes after reflecting off objects, enabling us to see them. He explained, too, that the picture formed not in the eye, but in the brain.

During the Arabic translation movement in Baghdad from the 8th to the 11th century, scribes translated philosophical and scientific works especially from Greek, but also the Persian, Syrian and Sanskrit languages.

The mathematician and astronomer al-Khwarizmi (c.780–c.850) is renowned for his book on algebra. His works had a great impact on the development of mathematics as we know it, for instance, we use Arabic numerals and the number zero, which originally came from India. The word "algorithm" derives from his name.

The alchemist and scientist Jabir ibn Hayyan (c.721–c.815) discovered a great many important processes still used in modern chemistry including the production of citric and acetic acid, distillation and crystallization. Jabir is known as the "father of Arab chemistry."

The Abbasid Caliphate

The fall of Alexandria in 641 marked the rise of the first of three great medieval Islamic superstates: the Rashidun, Umayyad and Abbasid caliphates. After the Abbasids toppled the Umayyad dynasty in 750, they ruled until 1258 over a Muslim world that, at its height, stretched from the Atlantic Ocean in the west to the mountainous borders of India and China. The Silk Road ran right across the region, ensuring prosperity and the spread of knowledge through this great variety of cultures. The Abbasid Caliphate has been called an Islamic golden age, during which the scholars of the caliphate took the lead in nearly every area of learning. They established the House of Wisdom in their capital, Baghdad, and both Islamic and non-Islamic scholars tried to collect and translate all the world's knowledge. Many ancient texts that would otherwise have been lost were translated into Arabic and Persian (and later into Turkish, Hebrew and Latin). When the Mongols sacked Baghdad in 1258, it was the beginning of the end of the caliphate.

Astronomy was important, not least for determining the times of day when Muslims were to pray, and to show the direction of the holy city Mecca.

The all-round genius al-Jazari (1136–1206) wrote a book detailing the workings of 100 "ingenious mechanical devices"—including a water clock atop an elephant statue, which he described as a fusion of Indian, African, Chinese, Persian, Greek and Islamic cultures.

Persian windmills had a vertical axis.

Europe was ravaged by raids by Vikings and robber gangs.

The Indian mathematician and astronomer Brahmagupta (c.598–c.668) was the first to write sums using negative numbers and the zero. Later he wrote an astronomy handbook for students.

In the early middle ages the three-field system replaced the two-field system. Farmers divided fields into three and rotated their crops: winter and summer grains and fallow fields were swapped around every three years.

The Chinese inventor Su Song (1020–1101) built a water tower 12 m / 40 ft high that also showed the position of the sun, moon and stars. Its bronze-cast mechanism drew on a thousand years of Chinese mechanical engineering.

By fitting a heavy plow with a coulter to cut the soil and wheels to bear the weight of its iron share, farmers could cut and turn heavier clay soil.

Plowing teams of oxen were replaced by faster, more nimble horses.

Adding a harness that put the weight on the horse's strong chest made much quicker and easier work of tilling soil.

The horseshoe was now in general use.

The early middle ages

In Europe, the early middle ages cover the period from the fall of Rome in 476 to the end of the 10th century. Although it was traditionally seen as a time of scientific standstill—the "dark ages"—this was not the case in agriculture. Thanks to the mild Gulf Stream climate and plenty of year-round rainfall, crop production in western Europe steadily increased, particularly with the advent of new techniques that helped speed up work on the land. Threats—and innovations—came from various invaders, including Vikings, Moors and bands of brigands.

The high middle ages

Europe's high medieval period lasted from around 1000 until 1250. Over roughly the same time period, harvests tripled in yield; because of this and a more varied diet, life expectancy increased and there was a population explosion. There was greater political stability in the 12th century and the economy and trade grew in step with the population.

The crusades were holy wars—waged, mostly, by Christians against Muslims of the eastern Mediterranean

The Chinese Song Dynasty circulated the world's first paper money in the 11th century.

Universities and cathedrals.

The working day began at sunrise and finished at sunset. The curfew bell rang in the evening as a signal to put out all hearths and fires, in order to avoid blazes in wooden buildings.

Water mills.

Stirrups had existed in China for a long time, but these now became important for heavily armed knights and cavalry.

The English bishop and scientist Robert Grosseteste (c.1175–1253) understood the importance of conducting experiments. He was at the forefront of the scientific method.

The German bishop, philosopher and scientist Albertus Magnus, also known as Saint Albert the Great (c.1200–80), introduced both Aristotelian and Arab ideas into European thinking. One of his students was the Italian philosopher and theologian Thomas Aquinas.

The 12th-century doctor Trota of Salerno in Italy became locally famous for her writings on treatments for women. After her death, however, she was practically erased from the history books for 800 years. In 2007, her authorship was reaffirmed.

Gunpowder, invented in China in the 9th century, led quickly to fire lances, which in turn led to cannons, bombs and firearms in the 13th century. Via the Mongols, these found their way to the Middle East, and so to Europe.

The English friar Roger Bacon (1220–94), also known as Doctor Mirabilis ("amazing"), was one of the earliest advocates of the scientific method (careful observation, rational questioning, and testing hypotheses through experiment). He found that firsthand experience led to understanding; an approach known as empiricism.

The medical school established in Salerno, Italy, in the 9th century held a collection of Greek, Arabic and European texts. It developed as the first and most important source of medical knowledge in western Europe. Women were also trained as doctors and professors of medicine.

Leather, metal and silk were used to make body protection for Japanese samurai.

It was partly thanks to Leonardo of Pisa (c.1170–c.1250), later known as Fibonacci, that the Hindu-Arabic number system became popular in Europe, in place of the more cumbersome Roman numerals. He is also remembered for the Fibonacci sequence: a series of numbers used to model growth in many disciplines.

The Fibonacci sequence starts with 0 and 1, then each digit that follows is the sum of the previous two digits.

13th century

If the early middle ages in Europe had seen the development of agriculture, the late middle ages (c.1270–1500) witnessed the steady rise of the medieval town. By the 13th century there were many large towns in Europe, some with more than 10,000 inhabitants. Townsfolk depended on farmers for food. Their lifestyle was based on exchange and trade, which would lead eventually to the rise of the wealthy middle classes.

The Polish philosopher Witelo wrote about nature, the universe and biology, but most of his writings were lost. His surviving work, Perspectiva (c.1274), which developed the ideas of the Arab scientist al-Haytham, shows that he was a profound thinker. He studied the geometry of light rays, the nature of color, perspective, and much more.

The Venetian trader and explorer Marco Polo (c.1254–1324) was not the first European to travel to China, but he was the first to write about his adventures there. He brought back from China the concept of paper money, and descriptions of coal and an intricate postal system, which some believe led to their widespread use in Europe.

Thomas Aquinas (1225–74) was the most influential theologian and philosophical thinker of the middle ages. He believed that knowledge could only be acquired if two sources were used: belief (theology) and reason (philosophy).

The French scholar Petrus Peregrinus wrote about his experiments with magnets. He identified their north and south poles and studied their practical use in compass needles.

Henrik Harpestrœng (c.1164–1244), a little-known Danish botanist, wrote a herbal consisting of 150 chapters on plants. Today, the book is an invaluable source of Danish medieval plant names.

No one quite knows when spectacles were first used. Spectacles were described in Italy around 1280, by which time they were probably also being used in China and the Middle East. Earlier still, the Roman emperor Nero is said to have used an emerald to view performances.

Marco

Polo

Madhava (c.1340–c.1425) of Kerala in India was regarded as one of the greatest mathematician–astronomers of the middle ages. He came up with pioneering contributions to the study of infinite sequences, analysis, trigonometry and algebra.

The cross-staff, a forerunner of the sextant, allowed users to measure distant objects and to fix the degree of latitude at their location by calculating the height of the sun.

This astrarium, or planetarium, showed the time and date, and the movement of the planets and sun. It was built by the Italian doctor, astronomer and engineer Giovanni Dondi dell'Orologio (c.1330–88), son of a clock-maker.

The Franciscan monk William of Ockham (1287–1347) was a philosopher and theologian. He is best known for the principle named after him, Occam's razor, which states that the simplest solution to any question is most likely the correct one.

Occam's razor is surprisingly useful: down the centuries, it has been applied to questions of physics, chemistry, biology, zoology, religion, ethics, philosophy, crime and punishment, and even computer design.

Popularized in Europe during the 12th century, the compass was in general use on ships by 1350. The oldest navigational compass—from the Chinese Han Dynasty, 1500 years earlier—was made from magnetized iron on a floating piece of wood. Later improvements included a bar magnet, better suspension and the compass rose.

The hourglass had been around for centuries, but it now came into everyday use to measure time. At sea, it stood up well to ship movement and was reasonably precise in determining longitude.

A bombard is the forerunner of the cannon.

The cannon was developed in Europe, initially for laying siege to defensive walls and breaking up massed enemy troops in battle. In China, where gunpowder originated, the generally nomadic peoples had less need for such cumbersome artillery.

The longbow is an ancient weapon. English and Welsh archers proved its worth against the French during the early battles of the Hundred Years' War, including Sluys (1340), Crécy (1346), Poitiers (1356) and, perhaps most famously, Agincourt (1415).

Thanks to the use of buttresses and vaulted ceilings, stonemasons were able to build ever-taller Gothic cathedrals, with thinner walls and large, stained glass windows.

Since Roman times, soldiers mainly wore a coat of mail for protection. In the 14th century, with the shortage of skilled manpower to craft mail, knights tended to wear wrought-iron or steel plate. After the 15th century, the use of such protective clothing declined with the development of more effective firearms and crossbows.

Gargoyle (water spout).

Pinnacle.

Hounskull ("dog-faced") bascinet.

Flyer.

Sabaton.

Wall pier.

Flying buttress pier.

A group of scientific thinkers, nearly all connected with Merton College at Oxford University, called themselves the "Oxford Calculators." Keenly interested in the philosophical or logical "why" of the nature of things, they made real advances in mathematics and physics.

14th century

In the 14th century, large parts of Europe had to cope with a wetter climate. This ruined crops and contributed to the great famine of 1315–17, in which millions died. Many farmers, desperate for food, flocked to the cities—but the poor living conditions in crowded cities encouraged sickness and disease. The plague, or Black Death, killed tens of millions of Europeans. Meanwhile, following the death of Emperor Charles IV of France in 1328, the Hundred Years' War was fought between England and France. To medieval Europeans, it seemed that God was exacting terrible revenge as the prosperity of the 13th century came to a grinding halt.

At the command of the Chinese emperor Yongle, the Yongle Dadian was compiled by 2000 scholars at the University of Nanking. At more than 11,000 volumes, it would remain the largest encyclopedia in the world for the next six centuries.

Aberdeen, Barcelona, Basel, Bordeaux, Caen, Copenhagen, Glasgow, Leipzig, Leuven, Mainz, Nantes, St Andrews, Tübingen, Turin, Uppsala and Valencia were among the 30 or so universities founded in Europe during the 15th century.

The Renaissance architect and engineer Filippo Brunelleschi (1377–1446) is best known for his design of the huge octagonal dome of Florence Cathedral, though he did not live to see it completed. This engineering marvel, built entirely without scaffolding, contains more than four million bricks.

The world's oldest working astronomical clock is found in Prague. Built in 1410, it displays the positions of the sun and moon, as well as telling the time.

Printing had been refined in China for several centuries before it became popular in the west in medieval times—and even then, Europeans were slow to use it widely. But with the advent of movable type in 1455, printed books caught on swiftly.

The introduction of the moveable-type printing press by German inventor and publisher Johannes Gutenberg (c.1397–1468) was one of the most influential events in world history.

The Korean engineer, inventor and scientist Jang Yeong-sil (c.1390–1442) invented rain and water gauges and made improvements to many existing inventions, such as the water clock and sundial.

15th century

From the 15th century to the mid 17th century, powerful countries increasingly sailed forth looking to enrich themselves, find new trading routes and knowledge, and to bring new land under their control through force. Shipbuilding, navigation and mapmaking all improved and Europeans used sea charts for the first time.

Though he wasn't the first to suggest it, the Polish astronomer Nicolaus Copernicus (1473–1543) was important in proposing a heliocentric model of our solar system. That is, he wrote that Earth orbited the sun. Before that, most scholars believed that the sun, stars and planets orbited Earth. His most famous work is De Revolutionibus Orbium Coelestium ("Concerning the turning of the heavenly spheres").

Brace and bit.

The Portuguese explorer Diogo Cão (c.1450–86) reached the mouth of the Congo River on Africa's west coast, and later sailed to Namibia. He realized that Africa was far bigger than Europeans had originally thought.

Arquebus.

The admiral Zheng He (c.1371–c.1435) led seven expeditions round the Indian Ocean to extend Chinese control over the region and search for treasure. His fleets contained several hundred ships; some were said to be much larger than any wooden vessel built before or since.

The European century of exploration began when Portuguese sailors looked for an alternative route to Asia. Even the Italian Christopher Columbus (c.1451–1506) was searching for Asia in 1492 when he came upon America instead—the Vikings had been to Newfoundland. Columbus's expedition wasn't the first European contact with America. Columbus died believing that he had found a new route to India.

Leonardo da Vinci (1452–1519) wrote and illustrated the 72-page Codex Leicester between 1504 and 1508. Using mirror-writing (from right to left), he noted his observations about architecture, mechanics, science, painting and the human body. Microsoft founder Bill Gates bought the manuscript in 1994 for USD30.8 million.

The Kõpu lighthouse on the Estonian island of Hiiumaa has been operating continuously since it was built in 1531... nearly half a millennium ago.

Andreas Vesalius (1514–64) was a Flemish doctor and surgeon. He brought about a revolution in medicine when he published the first complete handbook of human anatomy in 1543, De Humani Corporis Fabrica Libri Septem ("Seven books on the structure of the human body").

Bernardino de Sahagún (1499–1590) was a Spanish missionary in New Spain (today's Mexico). He wrote about Aztec culture in Spanish and in Nahuatl (an Aztec language) and is widely considered the first anthropologist.

This Aztec giant calendar stone was probably carved between 1502 and 1521 from volcanic rock. A central image of what appears to be the sun god is surrounded by animals and objects representing dates and eras in the creation story.

Michael Servetus, a Spanish doctor and theologian, described the circulation of blood to the lungs. But he was punished for questioning religious ideas of his day and was burnt alive—on a heap of his own books.

Peter Henlein (1485–1542), a German maker of clocks and clocks, invented the first proper watch in 1505. It was about the size and shape of a golf ball.

16th century

Cannons and muskets helped armies win battles but, while there were still knights and nobles, it was the royal treasuries especially that could afford the new form of warfare. This military revolution forced a radical change in strategy and tactics in 16th- and 17th-century Europe. It became a competition between enormous defence works and heavy artillery. Drill and discipline became important so that musketeers would learn to fire salvos in order and in time. Not even the wealthiest towns could finance these kinds of wars. Small towns vanished, affecting where power was concentrated and laying the foundations of the modern city.

Taqi ad-Din Muhammad ibn Ma'ruf (1526–85) was an Ottoman scientist and mechanician. Under his guidance a large observatory was built in Constantinople, where a giant armillary sphere enabled people to study the planets.

The Ming Dynasty scholar Li Shizhen (1518–93) wrote a compendium describing 1895 drugs from Chinese medicine, and containing 1100 illustrations and 11,000 prescriptions. This huge undertaking, with over 50 volumes, took him nearly 30 years to complete.

Tycho Brahe (1546–1601) was a Danish astronomer famous for his very accurate observations. His discoveries included a new star in the constellation Cassiopeia. We know now that it was a supernova.

To make maps suitable for seafaring, the Flemish cartographer and globe-maker Gerardus Mercator (1512–94) designed charts with a new projection that displayed compass directions accurately (although areas near the poles appeared distorted). He also introduced the word "atlas."

Swiss naturalist Conrad Gessner (1516–65) was the founder of modern zoology, naming many plant and animal species, and a compiler of encyclopedias.

The German astronomer Johannes Kepler (1571–1630) discovered that the planets in the solar system had elliptical, not circular, orbits. This updated the heliocentric model of Copernicus.

English mathematician John Wallis (1616–1703) was the first to use the lemniscate symbol ∞ to describe infinity.

The Italian engineer, physicist and astronomer Galileo Galilei (1564–1642) was one of the first to explore the heavens with a telescope. Galileo demonstrated that Copernicus had been right when he suggested that the planets orbited the sun.

In 1628 the British doctor William Harvey (1578–1657) published a book that, for the first time, fully described the circulation of blood through the heart and body.

English doctor Richard Lower (1631–91) demonstrated that blood transfusion was possible from one animal to another, and even from animal to human—a xenotransfusion.

In 1637 the French philosopher René Descartes (1596–1650) wrote his "discourse of a method for the well guiding of reason, and the discovery of truth in sciences," which contains the famous dictum "Cogito, ergo sum" (I think, therefore I am).

The discoveries of French physicist and mathematician Blaise Pascal (1623–62) included a law named after him: Pressure applied to a fluid in a completely full and closed container is transmitted equally to every point of the fluid and the walls of the container. Pascal also built one of the first mechanical calculators, the Pascaline.

Francis Bacon (1561–1626) was an English philosopher, statesman and author. His method of research was named the Baconian method or, more simply, the scientific method. It is a systematic way of gathering knowledge.

Louise Boursier (1563–1636) was a French midwife. Her guidelines about pregnancy and birth were translated into Latin, German, English and Dutch and were in use until the end of the century.

The Dutch inventor Cornelis Drebbel (1572–1633) designed a so-called perpetual motion clock and also the first microscope with two convex lenses. He even built the first navigable submarine, which was tested by King James I of England.

17th century

As new concepts and methods paved the way for modern science, this century was a time of inquiry, research and moving carefully ahead. Scientists often were polymaths: interested in many different fields of knowledge and thinking. Science based itself more on observation but did not fully abandon classical science, religion and superstition: for example, Tycho Brahe immersed himself in astronomy without giving up his astrological predictions, and Johannes Kepler continued to make horoscopes even while proposing his three laws of planetary motion.

The Irish philosopher and chemist Robert Boyle (1627–91) wrote what we now know as Boyle's law: The pressure of a given quantity of gas varies inversely with its volume at constant temperature.

Isaac Newton (1642–1727) played a leading role during the scientific revolution of the 17th century. He laid the foundations of scientific analysis, developed a theory of the scientific worldview and formulated the law of gravity. His work determined the color spectrum and his theory of relativity in the early 20th century until Albert Einstein published law of gravity.

Robert Hooke (1635–1703) was an English physicist who discovered the law of elasticity. His research ranged many branches of science; with his microscope, and he discovered structures in cork and plants that reminded him of the cells of monks. Hooke's in 1665 he published observations made

The Dutch physicist, mathematician and astronomer Christiaan Huygens (1629–95) built his own telescope and discovered Titan, Saturn's largest moon. Best known for inventing the pendulum clock, he also attempted to build a gunpowder motor.

Italian doctor, naturalist, biologist and poet Francesco Redi (1626–97) was the first person to challenge the theory of the spontaneous creation of life, by proving that maggots came from the eggs of flies (and not simply from rotting meat, as people thought.)

The English inventor and engineer Thomas Savery (c.1650–1715) invented the first commercially viable steam engine, a revolutionary method of pumping water. His fellow countryman Thomas Newcomen (1664–1729) did more work on the design and developed his own steam machine.

Dutch self-taught scientist Antonie van Leeuwenhoek (1632–1723) designed lenses and microscopes and made the first microscopic observations of muscle structure, bacteria, spermatozoa, and blood flow in small blood vessels.

In 1701 the English agricultural scientist Jethro Tull (1674–1741) invented a mechanical seed drill, which keeps seeds from being lost and wasted. This led to bigger harvests.

The German entomologist Maria Sibylla Merian (1647–1717) wrote and illustrated books based on firsthand observations of insects. Her Metamorphosis Insectorum Surinamensium ("The metamorphosis of the insects of Suriname," 1705) was influential in its description of butterfly life cycles.

The great winter of 1708/09 was the coldest in Europe for 500 years.

English soldier and trumpeter John Shore (1662–1752) invented the tuning fork in 1711.

The physicist and instrument-maker Daniel Gabriel Fahrenheit (1686–1736) invented the mercury thermometer in 1714 and gave his name to the first standardized temperature scale.

James Hargreaves' spinning jenny (1764), a spinning machine with several spools, was one of the most important advances in the 18th-century textile industry.

In 1707 English doctor John Floyer introduced counting the pulse over a minute's duration to check the heart rate.

Shuttle.

Johann Jacob Diesbach is credited with inventing Prussian blue—the first modern synthetic pigment—in Berlin around 1706. (It's also known as "Berlin blue.")

The invention of the shuttle, which carried the weft thread back and forth across the warp, enabled the mechanization of weaving looms, dramatically increasing the output of the textile mills.

Encyclopedias became popular during the Enlightenment, when people wanted to systematize knowledge and bring education to a wider public.

The German graphic artist Jacob Christoph Le Blon (1667–1741) came up with color printing using red, blue and yellow inks. Years later he added black, introducing the four-color printing process that is still in use today.

Hyla arborea (tree frog)

The Swedish botanist Carl Linnaeus (1707–78) devised the system of naming living organisms that we use today: this binomial (two-part) scientific nomenclature made it much easier to classify species.

18th century

One of the most important changes in the Western world was the Industrial Revolution, which had its beginnings in Great Britain. New agricultural techniques increased food production; the population doubled and the need for goods was great. Mechanization in the factories generated a change from people- and animal-generated power to other sources. That quickly led to a shortage of wood, which meant an increase in the use of coal, industry's first fossil fuel.

The first parachute (1783).

In 1792 the French inventor Claude Chappe (1763–1805) developed a nationwide network of semaphores: tall masts with moveable wooden arms. Staffed by observers, it was able to transmit letters and therefore messages, and was used in many countries as the first workable telecommunications system.

German astronomer Maria Margaretha Kirch (1670–1720) was famous for her work concerning the sun, Saturn, Venus and Jupiter—though, being female, she was denied a place in the Royal Academy of Sciences.

The British astronomer Edmond Halley (1656–1742) discovered proper motion: the movement of a star with respect to the background stars. Halley's comet is named after him.

Maharaja Jai Singh II commissioned the building of five observatories. The largest of these Jantar Mantar observatories was completed in 1738 in his capital city of Jaipur.

In 1783 the French Montgolfier brothers gave the first public demonstration of a hot-air balloon. They burnt straw and wool to heat the air. A later test flight carried a sheep, a duck and a rooster aloft.

The British astronomer William Herschel (1738–1822) discovered Uranus in 1781, the first sighting of a new planet since ancient times. He built well over 60 telescopes, including one with a length of 12 m / 40 ft.

France adopted the metric system in 1795. Today, this international system of weights and measures is used nearly everywhere in the world.

The first steamboat (1776).

One of the many inventions of American politician and scientist Benjamin Franklin (1706–90) was the lightning conductor (1749).

Framed spectacles (1727).

The Italian physicist Alessandro Volta (1745–1827) built the first electric battery in 1799. It was made from zinc and copper discs with either a salty or acidic solution in between, held together by glass rods. The volt (a unit of electric potential) is, of course, named after him.

The Scotsman James Watt (1736–1819) developed an efficient, modern version of the steam engine. It was used to drain water and to power machines.

The Turtle (1775) was the first military submarine. The American colonists hoped to use it during their war of independence from Britain to place mines near enemy ships, but technical difficulties sank this plan.

French priest and entomologist Pierre André Latreille (1762–1833) wrote a great deal about crustaceans and insects, which he divided up into families. As a priest, he was imprisoned for refusing to swear allegiance to the state. While in his cell he found a rare beetle and informed the prison doctor, who was so impressed he ensured Latreille's release.

Cumulus clouds appear woolly and bulging, with a clearly defined edge. They often have a flat, dark base.

Cirrus clouds are wispy streaks above Earth's surface formed by ice crystals.

Stratus clouds are shapeless grey bands, low to the ground.

In 1805 the Irish-born Rear Admiral Francis Beaufort standardized the wind force scale. In later life, he thoroughly modernized British naval charts.

Luke Howard (1772–1864), a British chemist and meteorologist, came up with the names we still use for the various kinds of clouds. There are four main cloud families, often used in combination: for instance, cumulonimbus is a puffy raincloud.

French naturalist Jean-Baptiste Lamarck (1744–1829) coined the word "invertebrates" to refer to animals without a backbone and was also one of the first to use the term "biology" in its modern sense. He is best remembered, however, for his early ideas about evolution.

The Cornish inventor Richard Trevithick (1771–1833) built the first steam locomotive in 1804. He used it to transport coal.

The modern safety pin.

Canned food.

In 1802, after spending 20 years testing the anesthetic effects of various herbal compounds, Japanese surgeon Hanaoka Seishū (1760–1835) conducted the first-known operation on a patient under general anesthetic.

A Faraday cage is made from electrically conductive material. It is used to block electrical fields and static discharges, such as lightning.

The French physicist, inventor and mathematician André-Marie Ampère (1775–1836) conducted pioneering research into electromagnetism and its uses. The ampere (a unit of electric current) is named after him.

The English scientist Michael Faraday (1791–1867) conducted experiments with electricity and metals. This was the first investigation into semiconductors, the electrical building blocks of our modern industrial world.

William Playfair (1759–1823) was a Scottish jack-of-all-trades (his many careers ranged from engineer and editor to secret agent) who invented various graphic devices, including the bar chart and pie chart, for displaying statistics.

In 1817 the German Karl Freiherr von Drais (1785–1851) introduced his running bike, the draisine. Before this, running bikes could only move forward. The draisine had a steering mechanism.

The Scotsman James Clerk Maxwell (1831–79) showed that light, magnetism and electricity are all part of the same phenomenon: electromagnetic radiation. He is remembered as one of the greatest physicists of the modern era.

In 1843 the American government gave Samuel Morse the job of building the first long-distance telegraph line, between Washington and Baltimore. Soon telegraph cables were running beneath the Atlantic Ocean between North America and Europe.

Palaeontologist Mary Anning (1799–1847) spent her life searching for fossils in the limestone cliffs of Lyme Regis, the so-called Jurassic Coast in the south of England. Her many important finds included ichthyosaur, plesiosaur and pterosaur remains.

Louis Daguerre (1787–1851) gave his name to an early photographic process, the daguerrotype (1839). Along with Nicéphore Niépce, with whom he worked, he is considered one of the inventors of photography.

Braille, a raised type to help blind people read and write, is named after Frenchman Louis Braille (1809–52), who published the system in 1829.

Frenchman Nicéphore Niépce (1765–1833) is credited with taking the first permanent photograph. The view of rooftops from his room in Le Gras is his best-known photo.

The Belgian musical instrument maker Antoine-Joseph "Adolphe" Sax (1814–94) patented the saxophone in 1846.

Ada Lovelace (1815–52) was an English mathematician and is often called the first computer programmer. She wrote programs for an early kind of computer designed by Charles Babbage.

American inventor and artist Samuel Morse (1791–1872) developed an electric telegraph in 1838 but he is best known for his Morse code (1835), a system of dots and dashes for spelling messages over the telegraph.

The first postage stamp was stuck onto a letter on 6 May 1849. Because of its appearance, it was called the Penny Black. Today, Penny Blacks are rare and very valuable.

The early 19th century

Industrialization spread rapidly from Great Britain to the rest of Europe and America. As a result of the economic shift from agriculture to production, many people moved to the new factories for work. Soon, more people lived in cities than in the countryside, leading to overcrowding. A transport network of trains and steamships developed to take raw materials and finished products to their destinations.

In 1851 French scientist Léon Foucault (1819–68) used a giant pendulum to prove that Earth turned on its axis. As the bob moved, it traced a line in the air. As the day wore on, the line appeared to change direction according to Earth's rotation.

The Swedish inventor Alfred Bernhard Nobel (1833–96) had 355 patents registered to his name. His best-known invention is dynamite, but he also established a set of prestigious awards, the Nobel Prizes.

The American Thomas Alva Edison (1847–1931) purchased patents of inventions that looked interesting and which he went on to improve. He brought people together so they could jointly work on and refine products, such as the phonograph, movie camera and light bulb.

The British nurse Florence Nightingale (1820–1910) laid the foundations for modern nursing.

German physicist Heinrich Hertz (1857–94) proved the existence of radio waves, among many other discoveries. The Hertz, a unit of frequency, is named after him. His daughter Mathilde was an important biologist and psychologist.

In his 1859 book On the Origin of Species, English naturalist Charles Darwin (1809–82) gave partial proof that evolution had occurred and explained how it worked. Although Darwin is widely credited with developing the theory of evolution and natural selection, fellow British naturalist Alfred Wallace (1823–1913) played a vital role.

The German Robert Bunsen (1811–99) and his colleague Peter Desaga (1812–79) developed the Bunsen burner in 1855. Their burner perfected a previous model and is still used in laboratories today.

The later 19th century

Banks, exchanges and insurance companies were established to control and increase the new prosperity of the later 19th century. Governments of the industrial countries began to regulate the markets, and they formed technologically advanced armies to protect their wealth. A lot of raw materials were needed to increase production, and the search for more raw materials was one of the causes of modern imperialism—the further invasion and colonization of other countries.

Nikola Tesla (1856–1943) was a Serbian-American inventor. His research into radio signals led to a range of ideas, such as wireless power and light, a radio-control boat, and communications. Sadly, his laboratory burnt down in 1895, and it was Marconi who, in 1896, acquired the patent for the wireless telegraph. In 1891 Tesla developed the Tesla coil, which converted electricity into artificial bolts of lightning and sparks.

The French biologist, microbiologist and chemist Louis Pasteur (1822–95), with his wife Marie (1826–1910), conducted research into vaccination and "pasteurization" (heat treatment). Their discoveries of the causes and effects of illnesses saved many lives.

The first traffic light appeared in London in 1868 to control busy coach traffic. It was gas-lit and manually operated with a lever.

The Home Insurance Building in Chicago, built in 1885, was the first building with a steel frame. The building had 10 (later, 12) floors and was regarded as the world's first skyscraper.

The inventions of American Margaret Knight (1838–1914) were many and varied. In 1868 she designed a machine that could fold paper into a bag with a flat base.

The Benz Patent Motorwagen, designed in 1885 by German engineer Karl Benz (1844–1929), had an internal-combustion engine and a differential gearbox. In 1888 Karl's wife, Bertha, made the first long-distance car journey (during which she invented brake lining).

Jesse Wilford Reno (1861–1947) developed the first working escalator, at Coney Island, New York.

The first recognizable periodic system, a table of chemical elements arranged according to atomic number, was published in 1869 by the Russian chemist Dmitri Mendeleev (1834–1907).

The Italian electrical engineer Guglielmo Marconi (1874–1937) investigated radio transmission over long distances. Various scientists were looking for ways to send and receive electrical signals using radio waves.

Swiss pocketknife.

In 1879 Thomas Edison improved the design of the electric light bulb. This reliable source of light was useful for night shifts in factories.

In 1888, the Scottish vet and inventor John Boyd Dunlop (1840–1921) applied for the patent for the inflatable tyre.

In Paris in 1895 the brothers Auguste and Louis Lumière screened their first film: a 46-second clip of workers leaving the Lumière factory in Lyon.

While many inventors had a hand in the development of the telephone, it was Alexander Graham Bell (1847–1922) who in 1876 won the first American patent for the device.

The German general and inventor Ferdinand von Zeppelin (1838–1917) made a test flight in his airship LZ1 in 1900. The first commercial flight was in 1910. Within four years the German Airship Travel Corporation had conducted 1600 Zeppelin flights, transporting 37,250 people.

In 1903 the American brothers Orville and Wilbur Wright made the world's first powered, controlled flight. It lasted 12 seconds and covered 37 m / 122 ft at Kitty Hawk. The final flight that same morning lasted 59 seconds over a distance of 260 m / 850 ft.

American inventor Mary Anderson (1866–1953) created the first effective windscreen wiper in 1903.

In 1901 the American inventor King C. Gilette (1855–1932) created a razor with disposable blades.

In 1901 British ornithologist Edmund Selous (1857–1934) published the book Bird Watching. He was a staunch champion of this bird-friendly study.

Marie Curie (1867–1934) was a Polish-French scientist who won two Nobel Prizes. Her discoveries included two radioactive elements: polonium and radium. Science, medicine and industry all quickly found important uses for these elements. She also successfully isolated radium from uraninite. For decades now, radium has been used to treat cancer.

First vacuum cleaners (1901).

The VACUUM CLEANING COMPANY
Booths Patents

The A, B and O blood groups were identified by Karl Landsteiner (1868–1943) in 1901, a discovery for which he received the 1930 Nobel Prize in Physiology or Medicine.

In 1905 Albert Einstein (1879–1955) completed his thesis and published four articles with groundbreaking insights that changed physics forever. And he was only just getting started!

A standardized design for the typewriter.

In 1907 the Russian Ivan Pavlov (1849–1936) showed that by using a trigger—a sound, an electric jolt, or some other stimulus—during feeding he could, in time, condition dogs to produce saliva before he dished up their food: the Pavlov reaction.

In 1907 Belgian-American chemist Leo Baekeland (1863–1944) developed a fully synthetic material from synthetic ingredients. It was essentially the first plastic, and he called it Bakelite.

In 1908 the Japanese chemist Kikunae Ikeda (1864–1936) identified the chemical basis for umami, one of the five primary tastes (the others are salty, sweet, sour and bitter).

A Visit to the Seaside (1908), directed by English film pioneer George Albert Smith (1864–1959), was the first short film in Kinemacolor—a process of running a black-and-white film through red and green filters.

Washing machine.

Bathing machine

Demand for rubber soared as cars and bicycles became more popular. Initially this came from trees, but in 1909, a team from the German Bayer laboratories produced the first synthetic rubber.

The American businessman Henry Ford (1863–1947) brought about an industrial revolution when, in 1908, he began using an assembly line to produce cars. Cars became more affordable and by 1927 there were more than fifteen million Ford Model Ts on the road.

1900–1909

All sorts of networks were becoming more common around the world—from the telegraph to railways, gas and water supplies, and sewerage systems. What was at first only to be found in a few large cities was now everywhere, and the world seemingly grew smaller as it became ever more globalized. New technologies, such as the telephone and electricity in houses and factories, stimulated still greater productivity during this second industrial revolution, which ended with the outbreak of the First World War.

In their three-volume work Principia Mathematica (1910–13), British mathematicians and philosophers Bertrand Russell (1872–1970) and Alfred North Whitehead (1861–1947) aimed to set out the foundations of mathematics in relation to logic.

American astronomer Henrietta Swan Leavitt (1868–1921) discovered the relationship between the brightness of variable stars (Cepheids) and the period in which their brightness fluctuates. This enabled future astronomers to determine distances in the universe.

In 1927 the Belgian priest and professor of physics Georges Lemaître (1894–1966) put forward his hypothesis for an expanding universe. He is also the founder of the Big Bang theory.

In 1910 Theodor Wulf (1868–1946), a German physicist, went up the Eiffel Tower in Paris with an electrometer, capturing the first real evidence of cosmic radiation.

In 1912 the Polish chemist Casimir Funk (1884–1967) identified vitamins. His aim was to ward off diseases, such as beri-beri and scurvy, caused by nutritional deficiencies.

The Curtiss Model D was the first plane to take off and land on a ship.

The Panama Canal cuts across a narrow strip of land in Panama, Central America. On its completion in 1914, it linked the Atlantic and Pacific oceans. Beforehand, ships had to sail all the way around Cape Horn, the southernmost tip of South America.

In 1911 Roald Amundsen (1872–1928) was the first person to reach the South Pole.

In 1913 the English metallurgist Harry Brearley (1871–1948) invented a rustproof steel while researching the wear and tear of weapons. It is now known as stainless steel.

German physicist Hans Geiger (1882–1945) and his student Walther Müller (1905–79) developed the Geiger counter, a device for measuring X-rays and radioactivity.

The International Solvay Institutes were established by Ernest Solvay (1838–1922), a Belgian industrialist and chemist, to gain greater insights into chemistry and physics without sidelining other branches of the sciences.

Frenchman Jean-Henri Fabre (1823–1915), who wrote dozens of best-selling books on insects, is renowned as one of the founders of modern entomology.

War generated great leaps in technology. Nations pumped enormous sums and maximum resources into the deadly arms race, although there were spin-off benefits for society (such as the stainless steel for guns which found many later uses).

The International Astronomical Union (IAU), established in 1919, promotes international cooperation and coordination in astronomy. It also approves the names of celestial bodies.

In 1919 the British pilots John Alcock and Arthur Brown were the first to fly non-stop over the Atlantic Ocean. In a little over 16 hours, they flew from Newfoundland to Ireland— a feat that won them the £10,000 prize put up by a national newspaper.

The New Zealand scientist Ernest Rutherford (1871–1937) developed a model of the atom. He discovered the existence of protons: subatomic particles with positive charges. In 1908 he won the Nobel Prize for Chemistry for his research into radioactivity.

1910—1919

European nations ruled about 85 per cent of all the land on Earth, including 90 per cent of Africa. When European powers took over other places, the people living there suffered greatly. Imperialism, growing armies and an arms race created a Europe of alliances and hostilities that ultimately led to the First World War (1914–18). Science and technology helped determine how armies fought. Before the war, sonar had been used to locate shoals of fish, but now it helped track down enemy submarines. Devices invented or developed during the course of the decade included lightweight machine guns, incendiary bombs, artillery shells, short- or long-range listening apparatus, flame-throwers, tanks, sea mines and poison gas. People began to equip planes with bombs. Such use of science and technology on the battlefield shows that not all technological innovations were used with an eye to prosperity and peace.

In 1921, the Swiss psychiatrist Hermann Rorschach (1884–1922) published a book of "psychodiagnostics," which included the Rorschach or ink blot test, used to examine personalities and thinking processes.

In 1926, the Scottish inventor John Logie Baird (1888–1946) demonstrated an early television set to a group of scientists.

The Yagi, or beam, antenna was invented by two professors, Shintaro Uda (1896–1976) and Hidetsugu Yagi (1886–1976), from the Tohoku University in Japan.

In 1922 the British archaeologist Howard Carter (1874–1939) came across the tomb of Tutankhamun in the Valley of the Kings in Egypt.

In 1926 American scientist Katharine Burr Blodgett (1898–1979) became the first woman to receive a doctorate in physics from the University of Cambridge. Later on, her inventions included non-reflective glass.

In 1922 insulin was used to treat diabetes for the first time.

INSULI

Zip.

Pop-up toaster.

Vibraphone.

The lie detector, or polygraph, was used from 1921 during police questioning. People thought that by measuring blood pressure, heart rate and breathing, they would know if a subject was lying. Its reliability was often brought into doubt.

1920–1929

The First World War was followed by a period of relative quiet and peace. Factories that had concentrated on products for the war made the transition to consumer goods. New (or improved) products and processes developed during this decade include magnetic tape (used in sound recording), adhesive tape and foam rubber. But the fragile economic prosperity of the Roaring Twenties came to an end with the collapse of the American stock exchange, triggering the Great Depression.

The first feature-length motion picture with sound was The Jazz Singer (1927). The end of the silent film was now in sight.

In 1927 the German-Austrian engineer Fritz Pfleumer (1881–1945) invented magnetic tape for use in sound recording. Five years later, it featured in one of the first tape recorders, made by AEG.

In 1929 the Russian scientist Konstantin Tsiolkovsky (1857–1935) wrote Space Rocket Trains, in which he described the principle of multi-stage rockets. He wrote more books about space travel and is regarded as the father of rocket science.

In 1926 the American engineer Robert Goddard (1882–1945) launched the first rocket using liquid fuel. The rocket, named Nell, flew up 12 m / 40 ft and landed just 56 m / 180 ft away in a cabbage patch, but it was a giant leap in rocket technology.

The first iron with adjustable temperature control (1926).

The first practical, usable iron lung dates from 1928. The device supported the breathing of people whose lungs didn't work well or at all.

Electric refrigerator.

In 1928 the Scottish scientist Alexander Fleming (1881–1955) discovered by chance that the Staphylococcus bacteria he was growing in a Petri dish were killed by the fungus also growing there. That led to the first development of an antibiotic: penicillin. For this discovery, Fleming received a Nobel Prize in 1945.

At 381 m / 1250 ft, the Empire State Building in New York City was the tallest building in the world when completed in 1931. That changed in 1973 with the construction of the Twin Towers of the World Trade Center.

British pilot Amy Johnson (1903–41) flew solo from England to Australia in 1930, while in 1932 the American Amelia Earhart (b. 1897, disappeared in 1937) was the first female pilot to fly solo across the Atlantic Ocean.

The Focke-Wulf Fw 61, the first working helicopter, made its first flight in 1936.

The first working parking meter was set up in Oklahoma, USA, in 1935.

In 1937 the Ukrainian-American scientist Theodosius Dobzhansky (1900–75) published Genetics and the Origin of Species, one of the most famous works of neo-Darwinism, in which he explained the link between the theory of evolution and genetics.

In 1931 the research laboratory DuPont developed neoprene, or polychloroprene, a synthetic rubber. Four years later DuPont would seek a patent for nylon, a synthetic material.

Alka-Seltzer, a fizzy tablet treating indigestion, inflammation and pain, entered the market in 1931.

Plop, plop, fizz, fizz

In 1936 Danish seismologist Inge Lehmann (1888–1993) discovered that Earth's core consists of a fluid outer core and a solid inner core.

The American businessman Henry F. Phillips (1889–1958) bought the patent for the cross-head screw that now bears his name (along with the Phillips screwdriver).

The shopping cart was invented in 1937 by Sylvan Goldman (1898–1984), owner of the American Humpty Dumpty supermarkets.

The Richter scale, introduced in 1935 by American seismologist Charles Francis Richter (1900–85) and his German-American colleague Beno Gutenberg (1889–1960), was the first scale of measurement to use a number to estimate the energy released by an earthquake or a seaquake.

The BBC was established in 1922. In 1936 the "Beeb" broadcast its first regular programmes.

1930—1939

Despite, or perhaps because of, the Great Depression, hope prevailed during the 1930s. International exhibitions showcased technological progress. Science and technology were seen as the way to a better society. Skyscrapers, planes, cars, and developments in physics and biology all seemed like good reasons to be optimistic for the future. There were important developments, too, in atomic physics and in synthetic materials: the use of polymers, for instance, such as plastics and nylon, promised a new world of cheap, plentiful, mass-produced goods.

The Heinkel He 178 of 1939 was the first plane to fly using a turbojet engine—essentially the first practical jet plane.

DDT (dichlorodiphenyltrichloroethane) was found to be an effective insecticide in 1939. It is now widely banned as a dangerous chemical.

First vaccine against yellow fever (1937).

Walt Disney's Snow White and the Seven Dwarfs (1937) was the first animated film that lasted longer than one hour. It was an instant hit.

The Leslie speaker (1941), a loudspeaker with built-in tremolo.

Built by Westinghouse, the robot Elektro was on display in 1939 at the New York World's Fair. A year later it received a canine companion, the robot dog Sparko.

Musician George Beauchamp (1899–1941) and engineer Adolph Rickenbacher (1886–1976) introduced the first commercial electric guitar to the market in 1932: the Rickenbacker A-22, also known as the Frying pan.

The Hammond organ, launched in 1935, was originally a cheap electromechanical alternative to the pipe organ, but was also used in jazz, rock and gospel.

The American chemist Roy J. Plunkett (1910–94) discovered polytetrafluoroethylene (PTFE), or Teflon, in 1938. The most "friction-free" of all plastics, Teflon gives a non-stick coating to cooking pans, among other uses.

People lost trust in Zeppelins after a series of accidents. The tipping point came after the airship Hindenburg exploded in 1937 upon arriving in New Jersey.

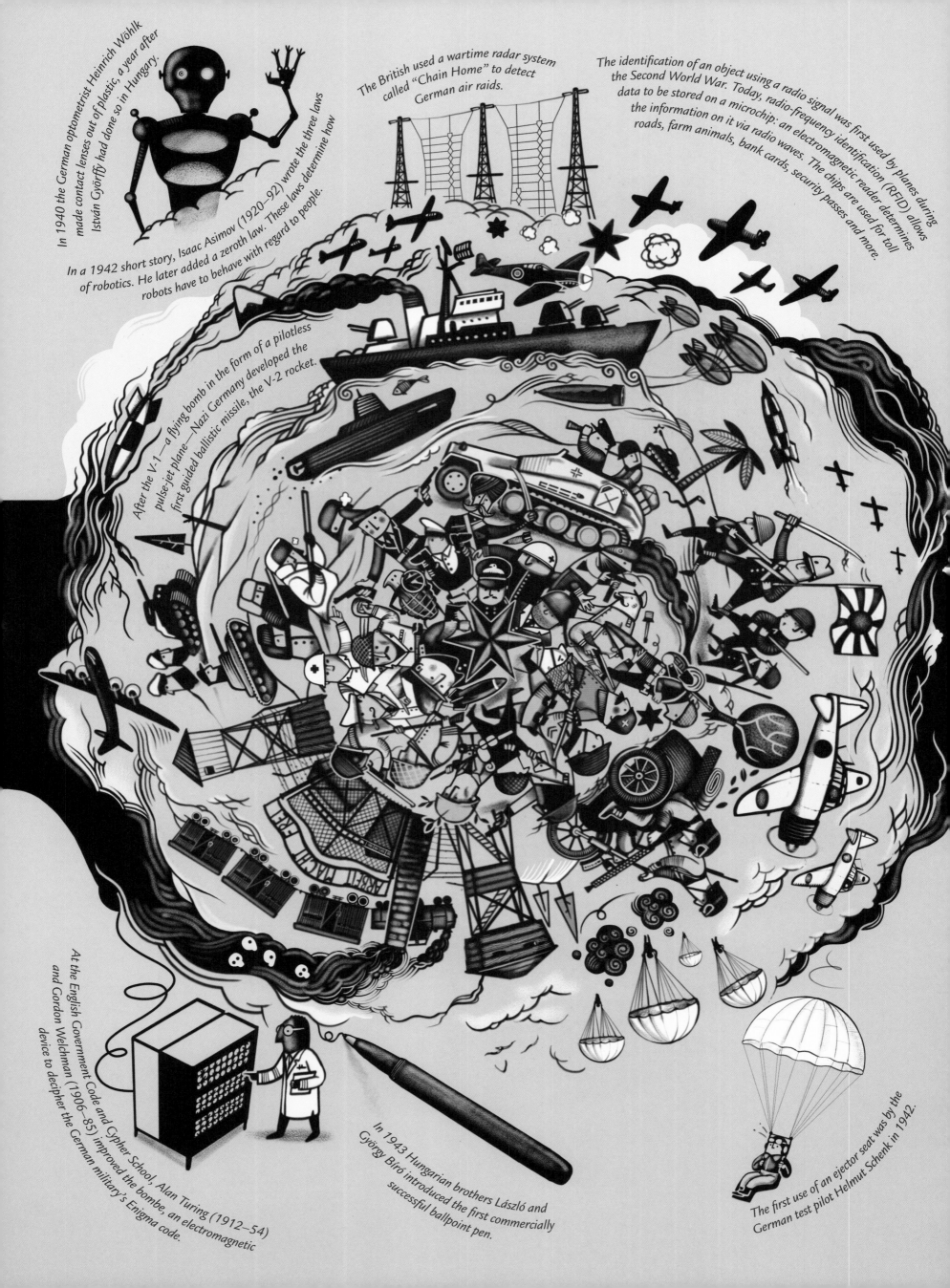

In 1940 the German optometrist Heinrich Wöhlk made contact lenses out of plastic, a year after István Györffy had done so in Hungary.

In a 1942 short story, Isaac Asimov (1920–92) wrote the three laws of robotics. He later added a zeroth law. These laws determine how robots have to behave with regard to people.

The British used a wartime radar system called "Chain Home" to detect German air raids.

The identification of an object using a radio signal was first used by planes during the Second World War. Today, radio-frequency identification (RFID) allows data to be stored on a microchip: an electromagnetic reader determines the information on it via radio waves. The chips are used for toll roads, farm animals, bank cards, security passes and more.

After the V-1—a flying bomb in the form of a pilotless pulse-jet plane—Nazi Germany developed the first guided ballistic missile, the V-2 rocket.

At the English Government Code and Cypher School, Alan Turing (1912–54) and Gordon Welchman (1906–85) improved the bombe, an electromagnetic device to decipher the German military's Enigma code.

In 1943 Hungarian brothers László and György Bíró introduced the first commercially successful ballpoint pen.

The first use of an ejector seat was by the German test pilot Helmut Schenk in 1942.

The Allies ended the war with Japan in 1945 when they dropped two atomic bombs on the cities of Hiroshima and Nagasaki, utterly destroying them and killing as many as 220,000 inhabitants. Japan surrendered a few days later.

In 1947 the American military launched the first living creatures (fruit flies) into space (and back) on a V-2 rocket captured from Germany.

In 1949 Albert II, a rhesus monkey, became the first mammal in space, in the US-launched V-2 rocket.

American wartime research and development into nuclear weapons was called the Manhattan Project. At its peak, around 130,000 people worked on the project.

The transistor was invented at Bell Labs in the USA in 1947. This electrical component can be used as a switch or amplifier and replaced the fragile thermionic valve or electron tube.

In 1947 the Norwegian anthropologist Thor Heyerdahl (1914–2002) sailed 6980 km / 4300 mi in his homemade raft, the Kon-Tiki, from South America to French Polynesia. He did this to find out if the early inhabitants might have come into contact with one another.

In 1949 the discovery of C14 (radiocarbon) dating allowed people to date organic material as far back as about 60,000 years. The American Willard Libby (1908–80) received the Nobel Prize in Chemistry for his role in the discovery.

In 1945 the British science-fiction writer and inventor Arthur C. Clarke (1917–2008) imagined a communication system based on satellites in geostationary orbits (they remain above the same place on Earth).

The United Nations, founded in 1945, is the world's most powerful intergovernmental body. It established the World Health Organization in 1948.

1940—1949

The Second World War (1939–45) also drove technology at breakneck pace, with some of the century's most significant developments—radar, nuclear energy, advanced rockets, jet planes and atomic bombs—taking shape almost from scratch within the span of the conflict. With the testing of the first atomic bomb in New Mexico on 16 July 1945, the nuclear age began. The transistor, invented in 1947, transformed electronics. From the end of the 1940s, universities, the military, and trade and industry developed computers to digitize and automate calculations.

The first plastic bags for the collection of waste.

Walter Lincoln Hawkins (1911–92), the first African American to be admitted into the National Academy of Engineering, is known for his research into polymers (in 1956, for instance, he developed an undersea telephone cable made of stronger plastic). He also pioneered plastics recycling.

In 1953 the American scientist James Watson and his English colleague Francis Crick announced that they had discovered the structure of deoxyribonucleic acid (DNA). DNA molecules are found in every living cell and determine how a plant or animal looks and functions.

The Indian American physicist Narinder Singh Kapany (1926–) did much research into optical fibers. The idea was not new, but his innovations ensured a far better result.

Scientists made the first successful synthetic diamonds in 1955.

The popularization of single-lens reflex cameras.

The first commercial transistor radios.

The first successful organ transplants were carried out in the 1950s.

The American biologist Rachel Carson (1907–64) wrote the bestseller The Sea Around Us in 1951. Her prophetic book Silent Spring (1962) would later describe the dangers of pesticides in the food chain and in the environment.

The work of the English chemist Rosalind Franklin (1920–58) was critical to the discovery of the structure of DNA using X-rays.

1950—1959

The 1950s saw huge leaps forward in technology—in the invention of new polymers, for instance, and in greater computing power. Electronic equipment became increasingly miniaturized, and would continue to be so, thanks to the new silicon transistor from Bell Labs (1954) and the integrated circuit from Texas Instruments (1958). (An integrated circuit is a chip on which several circuits are brought together.) Although military technology fed the growing cold war between the Soviets and Americans, it also found peaceful uses: rockets enabled people to launch satellites and explore space, while atomic reactors promised a new, potentially limitless source of energy.

In October 1957 the Soviet Union used a rocket to launch the first spacecraft, Sputnik I, into orbit around Earth. A month later they launched Sputnik II with a dog, Laika, on board.

In 1954 the Obninsk nuclear power station, near Moscow, was the first nuclear reactor in the world to be connected to the electricity network.

Commercial passenger flights became popular.

Jukeboxes were especially popular from the 1940s until the mid-1960s.

American physicist William Higinbotham (1910–94) created one of the first computer games in 1958. As part of an exhibition, two people could play tennis on an oscilloscope.

Televisions, already on the market since the 1940s, became popular in the 1950s.

In 1955 the National Physical Laboratory in England built the first accurate atomic clock.

In 1959 the Chinese American biologist Min Chueh Chang (1908–91) was the first to breed mammals—rabbits—through in vitro (test tube) fertilization. He also helped develop the birth-control pill.

The first hard disk, invented in 1956 by IBM (International Business Machines Corporation), had a capacity of 3.75 MB.

German inventor Artur Fischer (1919–2016), who came up with the first plastic wall anchor in 1958, has more than 1100 patents registered to his name, including the camera flashbulb and Fischertechnik toys.

The Clifford J. Rogers was the world's first purpose-built container ship, able to transport 600 containers. A container is a standard-sized box, carried on ships, trains and trucks. Containers were developed in the 1930s, but entered widespread use after the war.

Lego, beloved by young engineers everywhere, was patented in 1958.

STUFF

In 1961, during the Russian space mission Vostok 1, Yuri Gagarin (1934–68) was the first person launched into space. The cosmonaut circled Earth in less than two hours. He returned, safely, to world fame.

The communications satellite Telstar 1 (1962) was a privately sponsored space launch that relayed radio, telegraph, telephone and television signals.

The Unimate 1900 was the first industrial robot arm, used on an assembly line at General Motors in 1961.

Starting in 1960, the British primatologist and anthropologist Jane Goodall (1934–) studied chimpanzees in their natural environment in Tanzania. She discovered that they make use of tools. Her research changed the way these apes were studied and understood.

Working at DuPont, the American chemist Stephanie Kwolek (1923–2014) invented Kevlar—a heat-resistant, high-strength polymer—in 1964.

The first laser was built in 1960 by Theodore H. Maiman (1927–2007) in the Hughes Research Laboratories. He based it on the theoretical work of Charles Townes and Arthur Schawlow.

A bullet train first ran in Japan in 1964. The high-speed train could travel at 200 kmph / 125 mph, taking passengers from Tokyo to Osaka in just four hours. There is now a network of high-speed lines in Japan, called the Shinkansen.

Even though the microwave was invented in the 1940s, it only became small and cheap enough for household use in the mid 1960s.

The SI (International System of Units) was adopted in 1960. It is an expansion and modernization of the metric system. New, more precise definitions were needed to meet the needs of 20th-century science. Today the meter is defined as the distance light travels in a vacuum during a period of $\frac{1}{299,792,458}$ of a second.

In 1960 the Swiss bathyscaphe Trieste set a depth record of 10,916 m / 35,814 ft in the Mariana Trench in the Pacific, the deepest known place in any of the world's oceans.

The computer mouse was developed in 1963–64 by Douglas Engelbart and William English at the Stanford Research Institute, a physics think tank near San Francisco.

1960—1969

The Soviets had taken an early lead in the space saga with the launch of Sputnik, and in 1961 they showed they were still a step ahead when cosmonaut Yuri Gagarin became the first person in space. Less than 10 years later, in 1969, the astronaut Neil Armstrong planted the flag for America when he became the first to set foot on the moon. In the years since, several other countries have taken up the space challenge and more than 5000 spacecraft have been launched, from weather satellites and space probes to crewed spaceships and space stations.

Apollo 11 was the space flight that first took people to the moon. Neil Armstrong (1930–2012) and pilot Buzz Aldrin (1930–) landed the Apollo Lunar Module in July 1969. Six hours after landing, Armstrong set foot on the lunar surface. Since then, another 10 astronauts have walked on the moon.

The South African heart surgeon Christiaan Barnard (1922–2001) carried out the first successful heart transplant in 1967.

The American electrical engineer Marcian "Ted" Hoff (1937–) is a co-inventor of the first microprocessor, the Intel 4004.

The first cryonaut was James Bedford, who was frozen after his death in 1967. He hoped that in the future he could be brought back to life and that future science could then heal him.

The French explorer and researcher Jacques Cousteau (1910–97) studied the sea and all its life forms. In 1943 he developed the Aqua-Lung (or SCUBA) allowing people to dive for longer, and in 1959 a two-person submarine. In the 1960s he tested his Conshelfs: a series of seabed stations where "oceanauts" could live and study.

In the 1960s the American government, businesses and universities worked side by side to create a system in which computers in the US could exchange information. In 1969 they created an early form of the internet: ARPANET.

Mariner 9 was the first spacecraft to orbit another planet. It photographed the Martian surface in 1971–72, revealing volcanic activity and evidence that, long ago, water existed on Mars.

The space race between the US and the Soviet Union ended in 1975. In a bond of cooperation the American Apollo and the Russian Soyuz were linked together while orbiting Earth.

Willis Tower, or Sears Tower, in Chicago became the world's tallest building when completed in 1974, and remained so for 24 years.

The first LCD (liquid crystal display) watch, 1972.

In the mid-1970s many groups of computers were networked. Routers were invented to join the networks to one another. The original ARPANET grew into the internet.

The VCR (video cassette recorder) format was sold for domestic use from 1972.

The Godfather

Atari brought out the first video game, Pong, in 1972.

Barcodes became widely used.

9789401 450157

The first e-mail was sent in 1971, on the ARPANET network, by American computer engineer Ray Tomlinson (1941–2016). He addressed it to himself.

Art Fry created the Post-it Note in 1974, using the low-stick adhesive formulated six years earlier by his colleague Spencer Silver.

The first telephone conversation using a wireless device (1973).

The first humanoid robot in the WABOT series was completed in 1972 by Waseda University in Japan. It could walk, talk, watch and listen.

1970–1979

During the 1970s, the advent of the personal computer introduced people to the video game console and the first arcade video games. In business life, the organization and storage of data became ever more important and paper-based archives were digitized. The global oil crisis early in the decade had a major impact on daily life: shortages resulted in a huge increase in the price of oil, leading people to abandon "gas guzzlers" for smaller cars, and there was greater investment in solar, atomic and wind energy research.

In 1975 Bill Gates (1955–) and Paul Allen (1953–2018) set up the computer and software company Microsoft in New Mexico, USA. Twenty years later, Bill Gates was the richest man in the world.

The supersonic passenger plane Concorde made its first commercial flight in 1976.

In 1977 the American biophysicist Raymond Damadian (1936–) devised MRI (magnetic resonance imaging), a scanning technology used to map the body and certain body processes.

With Sony's first Walkman in 1979, people could listen to music anywhere and at any time.

Englishwoman Louise Brown (1978–) is history's first "test-tube baby." Using in vitro fertilization (IVF), egg cells are fertilized with sperm cells outside the body, and the resulting embryos are implanted into the womb.

Data cassette tapes became common.

In 1976 Apple released its first product, the Apple 1. This simple personal computer was the first to combine a keyboard with a microprocessor and a monitor.

The Wow! signal was a strong radio signal received by the Big Ear radio telescope in 1977. It appeared to be of extraterrestrial origin. A researcher wrote "Wow!" in the margin of a printout of the signal, hence the name.

In 1977 the American Voyager project launched two robotic probes to study the outermost solar system. Their mission has been extended three times, and Voyager 1 and Voyager 2 are still in interstellar space and continue to send back data.

In 1981 the United States launched the Space Shuttle, the first (partly) reusable spacecraft. Five shuttle craft undertook 135 missions up to 2011 when the fleet was retired.

Walkmans and boom boxes, invented at the end of the 1970s, were very popular and had a profound influence on the music industry and on youth culture.

In 1983 the domain name system (DNS) was introduced as the internet naming protocol.

The first CD (compact disc) and player came onto the market in 1982.

The development of MIDI (Musical Instrument Digital Interface) made it easier to integrate and synchronize electronic instruments, such as synthesizers and drum computers, into music.

Video camera.

Apple Macintosh.

IBM 5150.

The 1980s saw the first cases of AIDS, an illness caused by the human immunodeficiency virus (HIV). Millions of people have since died from it, but today there are treatments that keep HIV under control.

The Commodore 64 from 1982 was the best-selling computer of all time. Millions were sold.

Nintendo NES games computer.

Disk drive for the 5¼-inch floppy disk.

Using scanning tunneling microscopy (STM) people were able to image individual atoms for the first time. The developers Gerd Binnig (1947–) and Heinrich Rohrer (1933–2013) won the 1986 Nobel Prize in Physics.

The American ophthalmologist and inventor Patricia Bath (1942–2019) developed laser eye surgery to heal cataracts.

1980—1989

The digital revolution of the 1980s is often referred to as the third industrial revolution (following the first of the 1800–1900s and the second of the industrialized postwar era). Its impact was enormous. Our everyday life today is largely determined by the innovations and inventions developed in that decade when the personal computer, the internet, computer special effects and video games all came of age.

The military satellite Global Positioning System (GPS) was made available to the public during the 1980s.

In 1989 the New Zealand Department of Conservation began a recovery plan to save the kākāpō from extinction. The kākāpō is a giant, flightless parrot that can live for as long as 90 years.

In 1986 Halley's comet was studied by the "Halley Armada," consisting of five probes. One was from the European Space Agency, two were Japanese probes and two were jointly developed by the Soviet Union and France.

In 1988 the British physicist, cosmologist and scientist Stephen Hawking (1942–2018) published the best-selling book A Brief History of Time, in which he described the universe.

3D printers can print objects from a digital file.

In 1988 Nikon introduced the first commercial DSLR (digital single-lens reflex) camera.

In 1986 the nuclear reactor at Chernobyl exploded, releasing large quantities of toxic radioactive materials. The area is still uninhabitable.

Photoshop, an image-editing program from Adobe, was developed in 1988.

Pixar Animation Studios was established in 1986.

The first MS-DOS computer virus, named Brain, dates from 1986.

The computer game Tetris (1984).

The American biotechnological business Advance Genetic Studies was one of the first companies to modify a plant cell (1983) and one of the first to carry out field trials using genetically modified crops (1987).

Tim Berners-Lee (1955–), a British software developer, devised hypertext. Hypertext can link to other pieces of text or to documents with hyperlinks. It became the basis of the World Wide Web, invented by Berners-Lee in 1989.

The "family portrait" or "portrait of the planets" is a composite picture of our solar system using 60 photographs taken by Voyager 1 in 1990.

In 1990 the American space telescope Hubble was put into orbit around Earth. A telescope is able to take better images in space, beyond the gases of our planet's atmosphere.

In 1997 the American spacecraft Mars Pathfinder landed on the planet Mars with an exploratory robot, Sojourner, onboard.

In 1993 Shuji Nakamura (1954–), working with Isamu Akasaki and Hiroshi Amano, invented the bright blue LED, which made energy-saving white lighting possible.

The first VR (virtual reality) headsets came onto the market in the 1990s. VR is a simulated environment that resembles the real world.

The Tamagotchi, brainchild of Aki Maita (1967–) and Akihiro Yokoi (1955–), is a virtual handheld pet.

The American online shop Amazon sold its first book in 1995.

When commercial internet providers were allowed to sell individual internet connections for the first time in 1993, the use of the network grew enormously.

The first SMS (short media service) message was sent in 1992.

'Merry Christmas'

QR code (1994).

PlayStation consoles were brought out in Japan in 1994.

STUFF

Trilobite, the first commercial robot vacuum cleaner (1996).

With the completion of the Channel Tunnel underwater rail link between England and France in 1994, trains could travel between the two countries in 35 minutes.

The first module of the International Space Station was launched in 1998. A joint project between five nations, the station has been continually expanded since then.

In 1999 Bertrand Piccard (1958–) and Brian Jones (1947–) took nearly 20 days to travel round the world in a hot-air balloon... just for the fun of it.

In 1996 scientists proposed to clone the first adult mammal. The genetic material came from the udder tissue of a sheep. The Scottish scientists who carried out the experiment named the cloned sheep Dolly, after the country singer Dolly Parton.

Bluetooth, launched in 1999, is an open standard for wireless connections between devices at short distances. The logo is made up of runic letters representing H and B, initials of the 10th-century Viking leader Harald Bluetooth, who united Denmark and Norway.

AIBO is a robotic dog made by Sony since 1999 as an animal companion.

The first smartphones.

Google (1998).

WiFi

Devices can connect with each other and with the internet using the wireless connection Wi-Fi.

The Japanese designer Shigetaka Kurita (1972–) invented the emoji in 1999. Kurita found his inspiration for his first 176 emojis in Japanese comics, Chinese symbols and street signs.

In 1996 the IBM chess computer Deep Blue beat chess champion Garry Kasparov. Although Kasparov won the overall series, he lost the follow-up series to an upgraded Deep Blue in 1997.

Y2K! In 1999 everyone was worried that as soon as the year 2000 began many computer programs would crash, because the date operated with two numbers (for example, 92 instead of 1992). Many critical systems were adjusted beforehand and the worst was avoided.

In 1999 the Dutch inventor Theo Tempels (1938–) came up with the idea of putting a notch in Dutch rusks to make them easier to unpack.

1990—1999

The internet grew enormously during the 1990s. By 1992 more than a million computers were connected with one another; domain names gave each site its unique and searchable address. But this wasn't just machines exchanging information with one another: by integrating audio and video, the internet had become a vital means of communication between people. E-commerce, too, grew in scale, with the launch of sites such as Amazon and eBay. Advances in computing power led to greater understandings in, for instance, genetics, astronomy and particle physics.

According to a study published in Nature in 2004, trees have a theoretical maximum height of 130 m / 427 ft before gravity restricts their growth—they are unable to lift water any higher than this.

Wikipedia, a multilingual internet encyclopedia, appeared in 2001. It is the largest encyclopedia to be created in the last 600 years.

The first video on the video-hosting service YouTube was uploaded in 2005.

The fossil of a 428 million-year-old Pneumodesmus was discovered in Scotland in 2004. As far as we know, it is the world's oldest creature to have lived on the mainland.

The online network Facebook originated in 2003. Co-founder Mark Zuckerberg (1984–) is still in charge of the business.

ASIMO is a robot made by Honda in 2000, named after writer and scientist Isaac Asimov.

Apple introduced the iPod in 2001, a portable device on which thousands of songs can be stored and played.

The oldest footprints outside Africa were found at Happisburgh on the Norfolk coast in England, in 2013. They date from the early Pleistocene, more than 800,000 years ago.

Twitter, established in 2006, enables users to share short messages, or tweets.

The Segway, a self-balancing, one-person two-wheeler, came onto the market in 2001.

Last seen in 2006, the western black rhino and northern white rhino were declared extinct in the wild in 2011. IVF trials, using a female southern white rhino as a surrogate mother, may yet rescue the northern white.

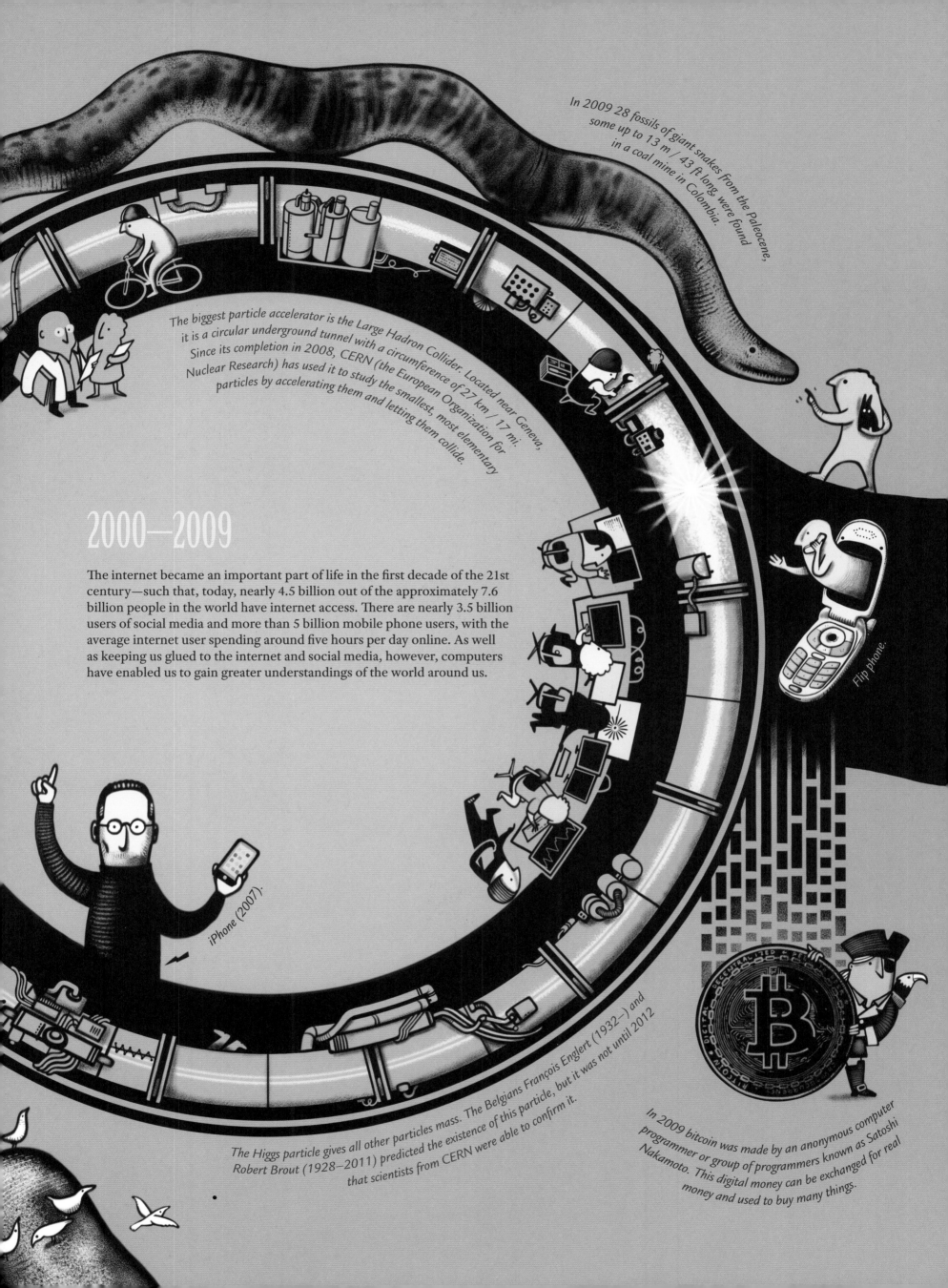

In 2009 28 fossils of giant snakes from the Paleocene, some up to 13 m / 43 ft long were found in a coal mine in Colombia.

The biggest particle accelerator is the Large Hadron Collider. Located near Geneva, it is a circular underground tunnel with a circumference of 27 km / 17 mi. Since its completion in 2008, CERN (the European Organization for Nuclear Research) has used it to study the smallest, most elementary particles by accelerating them and letting them collide.

2000—2009

The internet became an important part of life in the first decade of the 21st century—such that, today, nearly 4.5 billion out of the approximately 7.6 billion people in the world have internet access. There are nearly 3.5 billion users of social media and more than 5 billion mobile phone users, with the average internet user spending around five hours per day online. As well as keeping us glued to the internet and social media, however, computers have enabled us to gain greater understandings of the world around us.

Flip phone.

iPhone (2007).

The Higgs particle gives all other particles mass. The Belgians François Englert (1932–) and Robert Brout (1928–2011) predicted the existence of this particle, but it was not until 2012 that scientists from CERN were able to confirm it.

In 2009 bitcoin was made by an anonymous computer programmer or group of programmers known as Satoshi Nakamoto. This digital money can be exchanged for real money and used to buy many things.

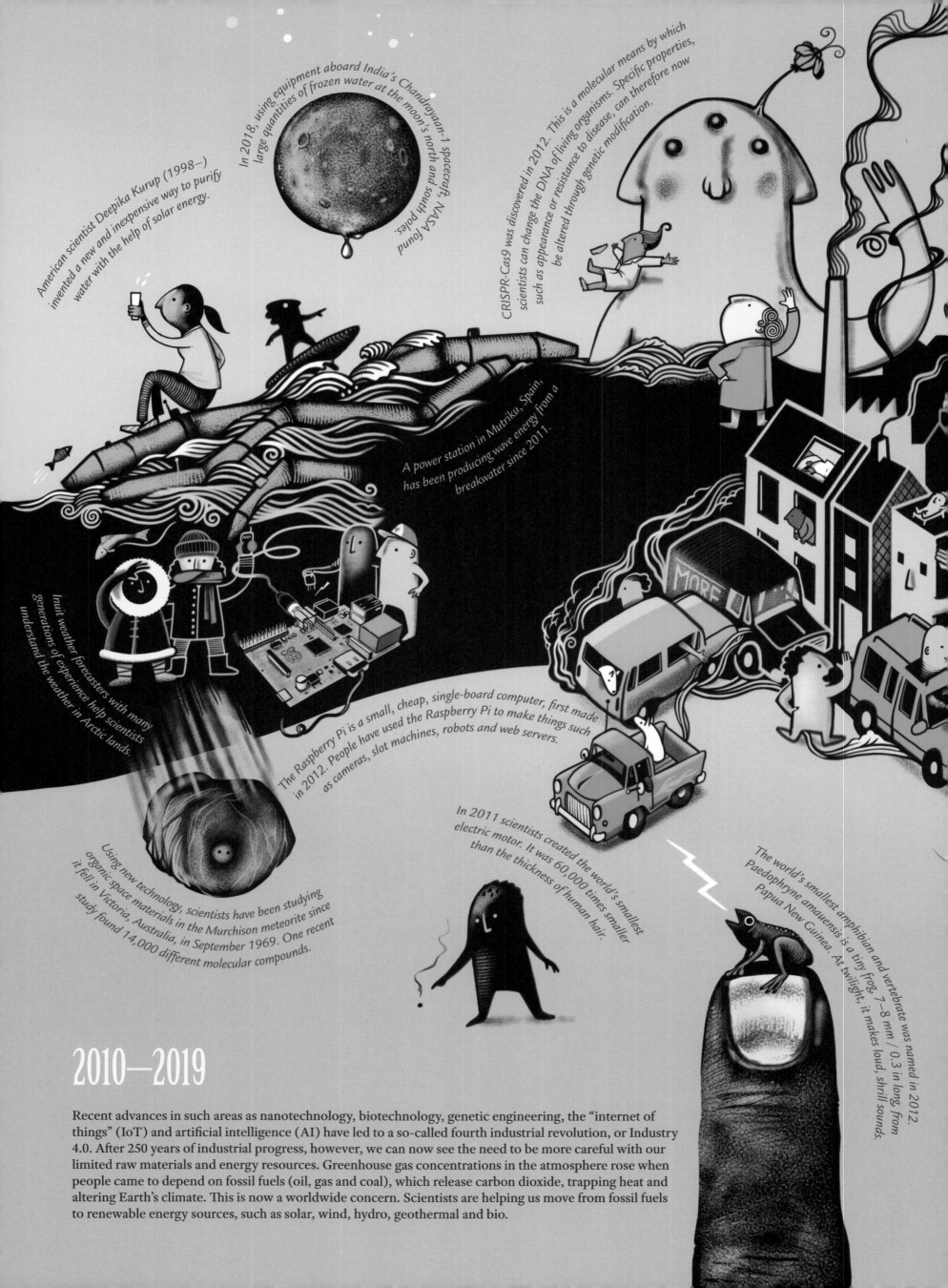

American scientist Deepika Kurup (1998–) invented a new and inexpensive way to purify water with the help of solar energy.

In 2018, using equipment aboard India's Chandrayaan-1 spacecraft, NASA found large quantities of frozen water at the moon's north and south poles.

CRISPR-Cas9 was discovered in 2012. This is a molecular means by which scientists can change the DNA of living organisms. Specific properties, such as appearance or resistance to disease, can therefore now be altered through genetic modification.

A power station in Mutriku, Spain, has been producing wave energy from a breakwater since 2011.

Inuit weather forecasters with many generations of experience help scientists understand the weather in Arctic lands.

The Raspberry Pi is a small, cheap, single-board computer, first made in 2012. People have used the Raspberry Pi to make things such as cameras, slot machines, robots and web servers.

Using new technology, scientists have been studying organic space materials in the Murchison meteorite since it fell in Victoria, Australia, in September 1969. One recent study found 14,000 different molecular compounds.

In 2011 scientists created the world's smallest electric motor. It was 60,000 times smaller than the thickness of human hair.

The world's smallest amphibian and vertebrate was named in 2012. Paedophryne amauensis is a tiny frog, 7–8 mm / 0.3 in long, from Papua New Guinea. At twilight, it makes loud, shrill sounds.

2010—2019

Recent advances in such areas as nanotechnology, biotechnology, genetic engineering, the "internet of things" (IoT) and artificial intelligence (AI) have led to a so-called fourth industrial revolution, or Industry 4.0. After 250 years of industrial progress, however, we can now see the need to be more careful with our limited raw materials and energy resources. Greenhouse gas concentrations in the atmosphere rose when people came to depend on fossil fuels (oil, gas and coal), which release carbon dioxide, trapping heat and altering Earth's climate. This is now a worldwide concern. Scientists are helping us move from fossil fuels to renewable energy sources, such as solar, wind, hydro, geothermal and bio.

In 2016 the Solar Impulse, a plane powered by solar energy, circumnavigated Earth.

Scientists cannot say how much warmer Earth will become. Most estimate that by the end of this century the temperature will have risen by 2–5 degrees celsius. The outcomes of a seriously altered climate include droughts, heat waves, fires, floods, hurricanes...

The space probe Chang'e 4, launched by the Chinese space agency (CNSA), landed on the far side of the moon in January 2019. It is named after the Chinese moon goddess.

In 2019 NASA discovered a four billion-year-old rock amongst moon samples collected on the 1971 Apollo 14 mission. Its great age shows that it came originally to the moon as a meteorite from Earth.

In 2018 scientists concluded that the extinct elephant bird of Madagascar, 3 m / 10 ft tall and weighing 800 kg / 1800 pd, was the biggest bird ever. It was probably behind the legend of the Roc, a bird that was once supposed to have eaten an elephant.

The Irish inventor Jane Ní Dhulchaointigh (1979–) won the 2018 European Inventor Award for the invention of Sugru, a moldable glue.

Online shopping was delivered by a drone for the first time.

In 2013 engineers managed to move small objects using sound waves.

The highly mobile robot Atlas, from Boston Dynamics, can make a backwards somersault.

In 2019 American and Chinese scientists enabled laboratory mice to see infrared light by injecting nanoparticles into their eyes.

A Maglev train reached a record speed of 603 kmph / 375 mph in 2015. Maglev trains do not ride on wheels but hover above the track; strong electromagnets propel them forward.

Despite progress in conservation, forests are still being felled. Trees feed on carbon dioxide. Fewer trees means less carbon dioxide is removed from the atmosphere.

Various countries are researching self-driving cars and, especially, electric cars. The American company Tesla is prominent in this field.

If completed, the Jeddah Tower in Saudi Arabia will, at 1008 m / 3308 ft, be the tallest building in the world.
Mars rovers are being equipped with robotic surveying helicopters.
Work has begun in Chile on the Giant Magellan Telescope, an assembly of seven primary mirrors. It should be operational by 2027 and will have a resolution 10 times greater than the Hubble space telescope.
Using advanced cloning techniques, Russia is planning to bring the extinct woolly mammoth back to life in order—in true Jurassic Park style—to breed the animals in Siberia.
The internet of things is steadily becoming a reality. Already there are now more smart devices connected to the internet and to each other than there are people. These "things" range from home fridges, cameras and thermostats to robots and self-driving cars.
The European Organization for Nuclear Research (CERN) wants to build a Future Circular Collider with a circumference of 100 km / 62 mi. The new research hub would be operational from 2040.
Artificial intelligence continues to develop. For example, great leaps have been made in the area of face recognition. People expect that technology will eventually allow machines to make diagnoses and treat illnesses, advise or influence consumers, detect crime…

2020—

Over the past century, rapid gains in science and technology have been made. International teams of experts are using specialized machines to push the boundaries of human knowledge. We are in search of the smallest particles and the farthest corners of the universe. Even as the climate crisis becomes perilous, we dream of a healthy Earth and of distant galaxies. Ever further, smaller, bigger and better.

This edition first published in 2020 by Gecko Press
PO Box 9335, Wellington 6141, New Zealand
info@geckopress.com

English-language edition © Gecko Press Ltd 2020
Translation © Bill Nagelkerke 2020
Original edition © Lannoo Publishers 2019
Original title: *Tijdlijn Wetenscap en Techniek*
Translated from the Dutch language
lannoo.com

This book was published with the support of Flanders Literature, flandersliterature.be

**FLANDERS
LITERATURE**

Edited by Matt Turner
Typesetting by Esther Chua
Printed in Italy

ISBN: 978-1-776573-00-4

petergoes.com
facebook.com/petergoesillustrator
instagram.com/goes.peter

For more curiously good books, visit geckopress.com